SIMPLIFIED DESIGN
OF MASONRY STRUCTURES

Other titles in the Parker–Ambrose Series of Simplified Design Guides

Harry Parker and James Ambrose
Simplified Design of Concrete Structures, 7th Edition

Harry Parker, John W. MacGuire and James Ambrose
Simplified Site Engineering, 2nd Edition

James Ambrose
Simplified Design of Building Foundations, 2nd Edition

James Ambrose and Dimitry Vergun
Simplified Design for Wind and Earthquakes Forces, 3rd Edition

Harry Parker and James Ambrose
Simplified Design of Steel Structures, 6th Edition

James Ambrose and Peter D. Brandow
Simplified Site Design

Harry Parker and James Ambrose
Simplified Mechanics and Strength of Materials, 5th Edition

Marc Schiler
Simplified Design of Building Lighting

James Patterson
Simplified Design for Building Fire Safety

James Ambrose
Simplified Engineering for Architects and Builders, 8th Edition

William Bobenhausen
Simplified Design of HVAC Systems

James Ambrose
Simplified Design of Wood Structures, 5th Edition

James Ambrose and Jeffrey E. Ollswang
Simplified Design for Building Sound Control

James Ambrose
Simplified Design of Building Structures, 3rd Edition

SIMPLIFIED DESIGN OF MASONRY STRUCTURES

JAMES AMBROSE

Formerly Professor of Architecture
University of Southern California
Los Angeles, California

JOHN WILEY & SONS, INC.

New York Chichester Weinheim Brisbane Singapore Toronto

This text is printed on acid-free paper.

To order books or for customer service please, call 1(800)-CALL-WILEY (225-5945).

This publication is designed to provide accurate and
authoritative information in regard to the subject
matter covered. It is sold with the understanding that
the publisher is not engaged in rendering legal, accounting,
or other professional services. If legal advice or other
expert assistance is required, the services of a competent
professional person should be sought.

Library of Congress Cataloging-in-Publication Data

Ambrose, James E.
  Simplified design of masonry structures / James Ambrose.
      p.   cm.
  Includes bibliographical references and index.
  I. Masonry. I. Title.
TH1199.A44 1991
693`.1--dc20   90-46401
  ISBN 0-471-52439-5 (cloth)
  ISBN 0-471-17988-4 (paper)

10   9 8 7 6 5 4

CONTENTS

PREFACE

This book presents the topic of design of masonry structures for buildings in a manner that is accessible to persons with limited training or experience in engineering investigation and design. It has been developed as a companion volume in the series of books developed by the late Professor Harry Parker and myself in the general topic area of architectural technology.

Topics treated here include the general use of materials, systems and details of construction, current building code and industry standards, and simple computations for ordinary structural elements of commonly used forms of masonry. Although the general topic of masonry extends to include a wide range of materials and applications, the work presented here deals principally with concerns for masonry as a structural material and not as a decorative finish material.

Although this book is not intended as a major source of data, examples of materials from various available sources are provided. These examples are discussed and, in many cases, used in example designs or investigations to illustrate their applications and to demonstrate their usefulness to the readers.

For readers who may use the book as a course text or for a self-study effort, study aids are provided at the end of the book. These consist of lists of significant words and terms, general questions, and exercise problems of computational form, all keyed to the individual chapters of the book. Answers to the questions and exercise problems are provided.

Structural masonry is used most extensively for elements that fundamentally resist compression, mostly as walls, columns, piers, or pedestals. Appendix A presents a summary of basic theories of the behavior of compression elements, which will be useful for readers with limited backgrounds in this topic.

Investigation of simple elements of structural masonry is still largely done by working-stress methods, using allowable stresses: for masonry without major steel reinforcement, this means the use of simple direct-stress formulas in the so-called empirical method. For reinforced masonry, formulas are largely derived by analogy with the working-stress method as applied to reinforced concrete. A digest of the working-stress method for reinforced concrete is provided in Appendix B.

In general, structural theories, computational formulas, and use of mathematics has been kept at a relatively simple level. Example designs cover most ordinary structural elements and systems and methods of construction in common use in the United States.

I am grateful to the International Conference of Building Officials, the American Concrete Institute, the National Concrete Masonry Association, and John Wiley & Sons for permission to use materials from their publications.

JAMES AMBROSE

Westlake Village, California
January 1991

SIMPLIFIED DESIGN
OF MASONRY STRUCTURES

1

INTRODUCTION

This book treats the topic of masonry construction that is intended for structural purposes in building construction, concentrating on the currently favored materials and forms of construction in the United States. This chapter presents discussions of the general usage of masonry for building construction, its historical development, and the various factors affecting its use for structural purposes.

1.1 MASONRY IN BUILDING CONSTRUCTION

The term *masonry* is quite broad, covering a considerable range of materials and forms of construction. Masonry using stones and bricks has deep roots, reaching back to ancient times in many cultures, with some basic uses preserved to present times.

The historical continuity of masonry construction, and its use for many well-known buildings in the past, make it very popular with the general public. Masonry is associated with a sense of permanency and durability, deriving from its use for many very old, well-preserved buildings as well as from its natural attribute of solidity.

It must be observed, however, that much of what appears to be masonry in modern construction is in all reality much different from the masonry construction of ancient times. Today, most buildings with marble, granite, and even brick exteriors are indeed frame structures of wood, steel, or reinforced concrete with thin veneers of the masonry materials. In fact, the masonry finish materials themselves may be imitations executed in plastic or fiber-reinforced cement.

1.2 STRUCTURAL MASONRY

Ancient builders would find it curious for someone to make a special designation of structural masonry. Their use of the mate-

1

rial was primarily for structures, although they often preserved the finest of their materials for the most prominent, visible locations. They also developed means for stretching the materials by creating structures with voids that were then filled with loose dirt, mud, and some forms of crude concrete.

Eventually, as civilizations progressed, the pressures to create buildings faster and more economically, in general, and the greater concern for labor costs resulted in shortcuts. This eventually led to techniques for making buildings of cheap construction appear to be of higher quality. Enter: the masonry veneer.

However, many of the attributes of masonry serve to preserve its use, even in these times. Its durability, fire resistance, and general solidity together with its structural potential make it a logical and popular choice in many situations. Structural masonry endures, although in many ways considerably different from that produced by ancient builders.

Structural masonry is presently used primarily for walls. In the past it was also used for foundations, abutments; and piers; in more advanced times, for arches, vaults, and domes. Today, construction with steel and reinforced concrete has largely displaced masonry, except for wall construction. This book deals primarily with the use of various forms of masonry for walls with structural purposes.

1.3 HISTORICAL DEVELOPMENT OF MASONRY

Masonry emerged in many primitive cultures when people simply made piles of rocks—for use as fortifications, retaining structures, dams, and, eventually, walls of buildings. Slowly the craft of making good rock piles was perfected and passed on to successive generations. The end of this continuing tradition produced some really elegant and imposing rock piles, such as the Egyptian pyramids, the Greek temples, the Roman aqueducts, the Gothic cathedrals, and the Great Wall of China.

Rocks were first used essentially as found in natural deposits. As experience grew and tools were developed, the rocks were increasingly used in some shaped form. Eventually stone masonry was perfected to an advanced craft, with massive structures formed as jigsaw puzzles of carefully shaped stones.

In cultures where stone was not readily available, or the crafts for quarrying it or shaping it were not developed, other forms of masonry were developed. Sundried and, later, fired-clay bricks were developed together with means for bonding them into a continuous mass. Natural materials were used to form mortar, crude concrete for filler, and plaster and stucco to protect the often soft masonry that had low resistance to weather or wear.

The legacy of thousands of years of construction experience remains with us in traditions of logically developed forms of masonry. More sophisticated, industrialized processes are used, but many of the details of present work with brick or stone masonry have direct roots in ancient constructions.

1.4 CURRENT USES OF MASONRY

Forms of structural masonry in use today in the United States have some traceable roots to ancient construction, but the materials, construction processes, and physical character of the work are quite different. For structural masonry the units most commonly used today are those consisting of fired clay bricks or precast concrete. These are produced in considerable variety, but are highly controlled, industrial products, available mostly in inventories of standard forms.

Mortar used for structural masonry is produced from packaged mixes and subject to standards set by building codes using criteria from nationally accepted, standards-writing agencies. Mortar for structural masonry is mostly made with portland cement, the same essential ingredient in structural concrete.

Because of its relative cost, construction using precast concrete masonry units (called CMUs)—or good old concrete blocks to the public—is widely used for structural purposes. Brick appears most often in veneered construction or in imitations with thin "brick" tiles adhered to a structural surface. Real brick walls, laid with mortar essentially as they were hundreds of years ago, are still used, but the labor-intensive construction is very expensive. If surface finishes are applied to the construction, structural masonry with CMUs is the definitive choice.

A broad distinction is made between masonry that is described as *reinforced* and that which is *unreinforced*. Reinforced masonry is made with steel bars in two directions embedded in the construction, in essence creating an emulation of reinforced concrete construction. If this is not done, the construction is classified as unreinforced, even though some steel reinforcing elements are generally used in all structural masonry. The specific conditions that establish this classification, and the limitations on usage of both types of construction are described more fully in Chapter 3.

In the masonry industry, the term *reinforcement* is mostly used to refer to steel elements incorporated into the masonry, by being embedded in mortar joints or placed in concrete-filled cavities in the construction. There are, however, many ways to structurally enhance the basic masonry construction by other means. These include the incorporation of stronger elements, such as precast concrete lintels, and form variations, such as the development of pilaster columns. Whereas reinforcing with steel is a modern development, many of the tricks developed by

ancient builders still have validity, the latter more so when the construction is that classified as unreinforced.

Masonry is often used for exterior walls, and a major concern of more recent times is the relative lack of resistance of solid masonry to thermal flow—of greater concern for heat loss in cold climates. For most occupancies, exterior masonry walls must be insulated in cold climates. Various forms of insulation are described in the discussions of the building design examples in Chapter 10.

Use of masonry construction, particularly for structures, is quite different in various regions of the United States. Concerns for cold climates include insulation (as just described), frost effects, and greater amounts of thermal expansion and contraction. Regions with high risk of windstorms or earthquakes frequently require the use of reinforced construction for all structural applications. This diversity in regional concerns has resulted in a considerable variation in forms of construction, commonly used materials, and the specific design standards established by local building codes. One result of this diversity is a profusion of industry organizations, as compared with the steel, concrete, and wood industries. There is no parallel to the form of dominance represented by the American Institute of Steel Construction (AISC) or the American Concrete Institute (ACI). The impact of this situation on use of design standards and the establishment of standard details for construction is discussed in the next two sections.

1.5 DESIGN AND CONSTRUCTION STANDARDS

As with any building construction, direct control of both design and construction practices is essentially in the hands of government agencies with jurisdiction for the building site. This control is exercised by enforcement of local building codes, zoning ordinances, and policies of the supervising agencies empowered to enforce the applicable ordinances. General concerns for building codes are discussed in Sec. 9.3.

For specific items, such as masonry construction, local codes may reflect some local concerns and experiences, but they usually use data and criteria from model building codes (such as the *Uniform Building Code* or the *BOCA Code*) or from standards-writing agencies. Despite regional differences, most codes use the same basic references for standards, so much of the technical data, design requirements, and construction requirements is similar in all codes. However, anyone doing actual design work is well advised to determine the code of legal jurisdiction for the work and to study its requirements carefully.

For this book, it was necessary to choose a single building code and a few references for the data and design criteria. Although other codes and some specific local problems may result in differences, most of the work shown here is quite common and generally applicable for the situations of usage described.

Building codes and industry standards are subject to continual revision as a result of changes in design and construction practices, emergence of new technology, results of research, and some experience gained from spectacular failures, such as those resulting from windstorms and earthquakes. Design for actual buildings must conform to the latest code requirements, and designers must be careful to ascertain the validity of any reference materials or design practices in this regard.

1.6   SOURCES OF DESIGN INFORMATION

Information about masonry construction is plentiful. There are two general groups

of such information. The first group is that dealing primarily with construction materials and processes. This is a useful source for details of the various types of construction. However, the information presented must be carefully evaluated for its proper application to specific situations. Depending primarily on the source of the information, the material presented may be highly regional in character, may be prejudiced in favor of certain materials, products, or priority processes, or may otherwise reflect some limited point of view of the publishers.

The second type of information is that dealing with the methods and procedures for structural investigation and design. This is often presented in industry standards or building codes with some degree of authority. Particular attention should be given to the building code that is legally applicable to any particular building project. Basic structural theories and analytical techniques do not change very rapidly, although the data used for design procedures must reflect the latest condition of the technology and prevailing construction practices.

Designers must establish their own sources of information. These must be as up to the minute as possible, appropriate to the region and nature of the work, and generally conforming to prevailing design and construction practices. Any published materials used as references must be carefully evaluated for reliability and general neutrality. Industry sources are indispensable, but one must be reasonable in appreciating the point of view of those who are in business to sell products of a particular type. You can probably rely on the brick manufactures to give you information on the proper use of bricks, but you cannot expect them to be neutral about the relative merits of brick construction versus other alternatives.

The sources for information used in developing the presentations in this book are listed in the References at the end of the text. These are all reliable sources, but there is no intention to advocate them as the best or only sources. Different sources may be more applicable for specific situations.

The type of information sources and the usefulness of the information they provide may be summarized as follows:

1. *Building Codes and Industry Standards*. These should generally be used as basic references, with special note of their timeliness and any legal enforcement.

2. *General Texts and Handbooks, Privately Published*. If not supported by some particular faction of the industry, these may have some reasonable neutrality, although the particular experiences or points of view of authors and editors may provide some slant to the materials.

3. *Industry-Generated Materials*.
   These are sometimes indispensable in terms of specific data for real products. However, they may often edge toward having a promotional and advertising mission that should be recognized.

1.7 STRUCTURAL COMPUTATIONS

The computational work in this book is simple and can be performed easily with a pocket calculator. Structural computations for the most part can be rounded off. Accuracy beyond the third place is seldom significant, and many results presented have been so rounded off. In lengthy computations, however, it is advisable to carry one or more places of accuracy beyond that desired for the final answer. For the most part, the work was performed on an eight-digit pocket calculator.

1.8 USE OF COMPUTERS

In most professional design firms, structural computations done for final design work are performed with computer-aided procedures. Many standard programs are available for routine work, and much of the necessary data is accessible from computer-retrievable sources. Many industry and professional organizations have software that can be purchased for ordinary design work.

Use of computer-aided methods permits faster accomplishment of tedious and complex investigations, more feasible study of alternatives, and design work that is interactive with that of others working on the same project. Many current design standards and codes have requirements and procedures that imply the use of computer-aided methods for practical design utilization.

The value of computer-aided methods increases with the level of complexity or sheer length of the computational work. For the most part, the work shown in this book is hardly worth doing with a computer. However, the purpose of this book is basically instructional, and the hand operation of the full computational process allows for more involvement in the problems and the operation of their solutions.

1.9 UNITS OF MEASUREMENT

At the time of preparation of this edition, the building industry in the United States is still in a state of confused transition from the use of English units (feet, pounds, etc.) to the metric-based system referred to as the SI units (for Système International). Although a complete phase-over to SI units seems inevitable, at the time of this writing, construction-materials and products suppliers in the United States are still resisting it. Consequently, most building codes and other widely used references are still in the old units. (The old system is now more appropriately called the U.S. system because England no longer uses it!) Although it results in some degree of clumsiness in the work, we have chosen to give the data and computations in this book in both units as much as is practicable. The technique is generally to perform the work in U.S. units and immediately follow it with the equivalent work in SI units enclosed in brackets [thus] for separation and identity.

Table 1.1 lists the standard units of measurement in the U.S. system with the abbreviations used in this work and a description of the type of the use in struc-

tural work. In similar form, Table 1.2 gives the corresponding units in the SI system. The conversion units used in shifting from one system to the other are given in Table 1.3.

For some of the work in this book, the units of measurement are not significant. What is required in such cases is simply to find a numerical answer. The visualization of the problem, the manipulation of the mathematical processes for the solution, and the quantification of the answer are not related to the specific units—only to their relative values. In such situations we have occasionally chosen not to present the work in dual units, to provide a less

Table 1.1 UNITS OF MEASUREMENT: U.S. SYSTEM

Name of Unit	Abbreviation	Use
Length		
Foot	ft	Large dimensions, building plans, beam spans
Inch	in.	Small dimensions, size of member cross sections
Area		
Square feet	ft^2	Large areas
Square inches	$in.^2$	Small areas, properties of cross sections
Volume		
Cubic feet	ft^3	Large volumes, quantities of materials
Cubic inches	$in.^3$	Small volumes
Force, Mass		
Pound	lb	Specific weight, force, load
Kip	k	1000 lb
Pounds per foot	lb/ft	Linear load (as on a beam)
Kips per foot	k/ft	Linear load (as on a beam)
Pounds per square foot	lb/ft^2, psf	Distributed load on a surface
Kips per square foot	k/ft^2, ksf	Distributed load on a surface
Pounds per cubic foot	lb/ft^3, pcf	Relative density, weight
Moment		
Foot-pounds	ft-lb	Rotational or bending moment
Inch-pounds	in.-lb	Rotational or bending moment
Kip-feet	k-ft	Rotational or bending moment
Kip-inches	k-in.	Rotational or bending moment
Stress		
Pounds per square foot	lb/ft^2, psf	Soil pressure
Pounds per square inch	$lb/in.^2$, psi	Stresses in structures
Kips per square foot	k/ft^2, ksf	Soil pressure
Kips per square inch	$k/in.^2$, ksi	Stresses in structures
Temperature		
Degree Fahrenheit	°F	Temperature

Table 1.2   UNITS OF MEASUREMENT: SI SYSTEM

Name of Unit	Abbreviation	Use
Length		
Meter	m	Large dimensions, building plans, beam spans
Millimeter	mm	Small dimensions, size of member cross sections
Area		
Square meters	m^2	Large areas
Square millimeters	mm^2	Small areas, properties of cross sections
Volume		
Cubic meters	m^3	Large volumes
Cubic millimeters	mm^3	Small volumes
Mass		
Kilogram	kg	Mass of materials (equivalent to weight in U.S. system)
Kilograms per cubic meter	kg/m^3	Density
Force (Load on Structures)		
Newton	N	Force or load
Kilonewton	kN	1000 newtons
Stress		
Pascal	Pa	Stress or pressure (1 pascal = $1 N/m^2$)
Kilopascal	kPa	1000 pascal
Megapascal	MPa	1,000,000 pascal
Gigapascal	GPa	1,000,000,000 pascal
Temperature		
Degree Celsius	°C	Temperature

Table 1.3   FACTORS FOR CONVERSION OF UNITS

To Convert from U.S. Units to SI Units, Multiply by:	U.S. Unit	SI Unit	To Convert from SI Units to U.S. Units, Multiply by:
25.4	in.	mm	0.03937
0.3048	ft	m	3.281
645.2	in.^{2}	mm^2	1.550×10^{-3}
16.39×10^3	in.^{3}	mm^3	61.02×10^{-6}
416.2×10^3	in.^{4}	mm^4	2.403×10^{-6}
0.09290	ft^2	m^2	10.76
0.02832	ft^3	m^3	35.31
0.4536	lb (mass)	kg	2.205
4.448	lb (force)	N	0.2248
4.448	kip (force)	kN	0.2248
1.356	ft-lb (moment)	N-m	0.7376
1.356	kip-ft (moment)	kN-m	0.7376
1.488	lb/ft (mass)	kg/m	0.6720
14.59	lb/ft (load)	N/m	0.06853
14.59	kips/ft (load)	kN/m	0.06853
6.895	psi (stress)	kPa	0.1450
6.895	ksi (stress)	MPa	0.1450
0.04788	psf (load or pressure)	kPa	20.93
47.88	ksf (load or pressure)	kPa	0.02093
16.02	pcf (density)	kg/m^3	0.06242
$0.566 \times (°F - 32)$	°F	°C	$(1.8 \times °C) + 32$

confusing illustration for the reader. Although this procedure may be allowed for the learning exercises in this book, the structural designer is generally advised to develop the habit of always indicating the units for any numerical answers in structural computations.

1.10   SYMBOLS

The following "shorthand" symbols are frequently used.

Symbol	Reading
$>$	is greater than
$<$	is less than
$\geq$	equal to or greater than
$\leq$	equal to or less than
$6'$	6 feet
$6''$	6 inches
Σ	the sum of
δL	change in L

1.11   NOMENCLATURE

Notation used in this book complies generally with that used in the 1988 ACI Code (Ref. 4) and the 1988 *Uniform Building Code* (Ref. 1). The following list includes all of the notation used in this book and is compiled and adapted from more extensive lists in the references.

A_c Area of concrete

A_e Net cross-sectional area of masonry

A_g Gross area, determined by outer dimensions

A_n Net area (same as A_e)

A_s Area of steel reinforcement

A'_s Area of compressive reinforcement in doubly reinforced section

A_v Area of shear reinforcement

C Compressive force

E_c Modulus of elasticity of concrete

E_m Modulus of elasticity of masonry

E_s Modulus of elasticity of steel

F_a Allowable compressive stress due to axial load only

F_b Allowable compressive stress due to bending

F_c Allowable stress in concrete (general)

F_s Allowable stress in steel reinforcement

F_{sc} Allowable compressive stress in column reinforcement (*UBC*)

F_y Specified yield stress of steel reinforcement (same as f_y)

I Moment of inertia

K Factor for resisting moment, masonry

M Bending moment

M_R Resisting moment of a reinforced member

N Axial load

P Concentrated load

P_a Allowable axial load for reinforced masonry column

R Factor for resisting moment, concrete

T Axial tension force

V Total shear force on a section

W (1) Total gravity load; (2) total horizontal wind load

a (1) Unit area; (2) height of compression zone in reinforced section, strength method

b Width of reinforced section

b_w Width of stem in T-beam

d Effective depth of reinforced section

e Eccentricity of a nonaxial load; distance from the centroid of the section to the point of application of the load

f_a Calculated compressive stress due to axial load

f_b Calculated compressive stress due to bending

f_c Calculated stress in concrete (general)

f'_c Specified compressive strength of concrete, psi

f_m Calculated compressive stress in masonry (general)

f'_m Specified compressive strength of masonry, psi

f_p Calculated bearing stress

f_s Calculated stress in steel reinforcement

f_v Calculated shear stress in concrete or masonry

f_y Specified yield stress of steel reinforcement

h Effective height (unbraced) of wall or column (same as h')

h' Same as h

j Factor for internal moment arm (jd) in reinforced flexural member

k Factor for depth of compressive stress zone (kd) in reinforced flexural member

l Length of span

n Modular ratio, E_s/E_c or E_s/E_m

p Percent of steel reinforcement, A_s/A_g

r Radius of gyration

s Spacing of reinforcement, center to center

t Thickness (overall) of concrete slab, masonry wall

v Unit shear stress, same as f_v

w Unit (lb/ft, etc.) of uniformly distributed load on a beam

ε Unit strain, total deformation divided by original length

ϕ Strength reduction factor (strength design)

Δ Change of; as in ΔL = change in length

2

MATERIALS FOR
MASONRY CONSTRUCTION

Because of its diversity, there is a considerable range of materials and elements used to produce masonry construction. This chapter presents discussions of the common elements of masonry construction with concentration on structural applications.

2.1 MASONRY UNITS

Masonry consists generally of a solid mass produced by bonding separate units. The traditional bonding material is mortar. The units include a range of materials, the common ones being the following:

Stone. These may be essentially natural form (called rubble or fieldstone) or may be cut to specified shape.

Brick. These vary from unfired, dried mud (adobe) to fired clay (kiln-baked) products. Form, color, and structural properties vary considerably.

Concrete Blocks (CMUs). Called *concrete masonry units*, these are produced from a range of types of material in a large number of form variations.

Clay Tile Blocks. Used widely in the past, these are hollow units similar to concrete blocks in form. They were used for many of the functions now performed by concrete blocks.

Gypsum Blocks. These are precast units of gypsum concrete, used mostly for nonstructural partitions.

The potential structural character of masonry depends greatly on the material and form of the units. From a material point of view, the high-fired clay products (brick and tile) are the strongest, producing very strong construction with proper mortar, a good arrangement of the units, and good construction craft and work in general. This is particularly important if the general class of the masonry construction is the traditional, unreinforced variety. Although some joint reinforcing is typical in all structural masonry these days, the term *reinforced masonry* is reserved for a class of construction in which major vertical and horizontal reinforcing is used, quite analogous to reinforced concrete construction.

Unreinforced masonry is still used extensively, both in relatively crude form (rough, "native" construction with fieldstone or adobe bricks) and in highly controlled form with industrially produced elements. However, where building codes are sophisticated and strictly enforced, its use is limited in regions with critical wind conditions or high seismic risk. This makes for a somewhat regional division of use, reinforced masonry being most generally used in southern and western portions of the United States, for example, while unreinforced masonry is widely used in the east and midwest.

With reinforced masonry, the masonry unit takes a somewhat secondary role in determining the structural integrity of the construction. This is discussed more thoroughly in Sec. 3.6.

Structural masonry consists most often of construction with concrete units. Units are available in a wide range of types and forms, with many finishes attainable for exposed construction. Construction faced with brick or stone is often backed up by a supporting structure with CMUs.

2.2   BASIC CONSTRUCTION AND TERMINOLOGY

As evolved to present times, masonry takes some ordinary forms that retain classic elements and terminology of ancient construction. Figure 2.1 shows some of the common elements of masonry construction. The terminology and details shown apply mostly to construction with bricks or concrete blocks.

Units are usually laid up in horizontal rows, called *courses*, and in vertical planes, called *wythes*. Very thick walls

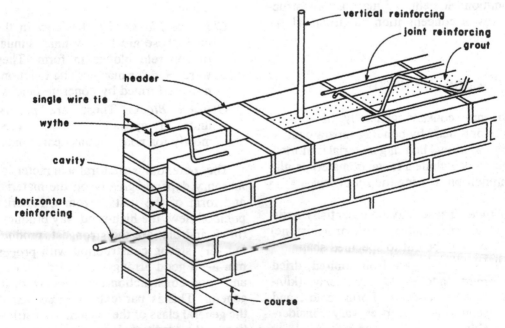

FIGURE 2.1. Elements of masonry construction.

may have several wythes, but most often walls of brick have two wythes and walls of concrete block are single wythe. If wythes are connected directly, the construction is called *solid*. If a space is left between wythes, as shown in the illustration, the wall is called a *cavity wall*. If the cavity is filled with concrete, it is called a *grouted cavity wall*.

The multiple-wythe wall must have the separate wythes bonded together in some fashion. If this is done with the masonry units, the overlapping unit is called a *header*. Various patterns of headers have produced some classic forms of arrangement of bricks in traditional masonry construction. For cavity walls, bonding is often done with metal ties, using single ties at intervals, or a continuous wire trussed element that provides both the tying of the wythes and some minimal horizontal reinforcing.

The continuous element labeled *joint reinforcing* in Fig. 2.1 is now commonly used in both brick and concrete block construction that is code-classified as unreinforced. For seriously reinforced masonry, the reinforcement consists of steel rods (the same as those used for reinforced concrete) that are placed at intervals both vertically and horizontally, and are encased in concrete poured into the wall cavities.

Unit dimensions may be set by the designer, but the sizes of industrially produced products such as bricks and concrete blocks are often controlled by industry standard practices. As shown in Fig. 2.2a, the three dimensions of a brick are the height and length of the exposed face and the width that produces the thickness of a single wythe. There is no single standard size brick, but most fall in a range close to that shown in the illustration.

Concrete blocks are produced in families of modular sizes. The size of block shown in Fig. 2.2b is one that is equivalent

to the 2×4 in wood—not the only size, but the most common. Concrete blocks have both nominal and actual dimensions. The nominal dimensions are used for designating the blocks and relate to modular layouts of building dimensions. Actual dimensions are based on the assumption of a mortar joint thickness of $\frac{3}{8}$ to $\frac{1}{2}$ in. (the sizes shown in Fig. 2.2b reflect the use of $\frac{3}{8}$-in. joints).

A construction sometimes used in unreinforced masonry is that of a single wythe of brick bonded to a single wythe of concrete block, as shown in Fig. 2.2c. In order to install the metal ties that affect the bonding, as well as to have the bricks and blocks come out even at the top of a wall, a special-height brick is sometimes used, based on either two or three bricks to one block.

Since transportation of large quantities of bricks or concrete blocks is difficult and costly, units used for a building are usually those obtainable on a local basis. Although some industry standardization exists, the type of locally produced products should be investigated for any design work.

When exposed to view, masonry units present two concerns: that for the face of the unit and the pattern of layout of the units. Patterns derive from unit shape and the need for unit bonding, if unit-bonded construction is used. Classic patterns were developed from these concerns, but other forms of construction, now more widely used, free the unit pattern somewhat. Nevertheless, classic patterns such as running bond, English bond, and so on, are still widely used. See discussion in Chapter 4.

Patterns also have some structural implications; indeed, the need for unit bonding was such a concern originally. For reinforced construction with concrete blocks, a major constraint is the need to align the voids in a vertical arrangement to facilitate installation of vertical bars. Gen-

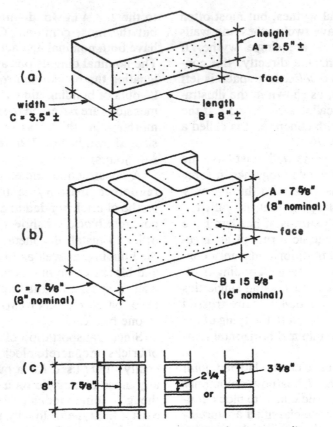

FIGURE 2.2. Dimensions and forms of masonry units.

erally, however, pattern as a structural issue is more critical for unreinforced masonry.

2.3 STONE

In most ancient cultures the earliest forms of masonry used stones as they were found. Stones were simply piled up to form crude structures. The craft of piling stones developed, and some simple lessons were learned about how to make stable, enduring piles. As the craft improved and tools were invented, the stones were shaped for better fit and ever more carefully developed stacks were made.

Mortar came later, possibly for various nonstructural reasons, such as keeping in-

sects and vermin from inhabiting the rock piles or to weather-seal walls against winter winds. Eventually, as better materials were discovered for mortars, their structural potential became more appreciated. Still, then as now, the best stone masonry achieved its basic strength and stability from the careful fitting and balancing of the stones.

Stones were usually too heavy to transport long distances, so even advanced societies had to rely on local resources. This resulted in many different forms and details of stone masonry, reflecting the limitations and potential of the available materials.

Piled stones can still be used for various purposes (see Fig. 2.3) and have a rough, natural-appearing character that is

FIGURE 2.3. Rock piles used as site structures; a very ancient heritage.

appealing. However, stone is mostly used for veneered construction at the present, often over frames instead of masonry backup structures. Some of the possibilities for use of stone are discussed in Chapter 6.

2.4   FIRED CLAY

A variety of masonry units have been developed from fired clay. The most familiar is the brick, but hollow blocks of fired clay were the predecessors of concrete blocks and are still in use for some purposes. Architectural terra-cotta was used extensively to imitate stone trim in the late nineteenth century and into the early twentieth century.

Unfired clay—essentially simple dried, hardened mud—was used for pottery and, where stone was not plentiful, for primitive brick construction. However, it was discovered that the burning of mud with a high clay content was found to produce a very hard, durable, and moisture-resisting material. With the development of mortar, brick construction was on its way.

Bricks were scorned as rough and inelegant by many early builders. Eventually, the craft and the materials were developed more finely, and many notable architectural achievements were produced with brick construction. Brick is a popular exterior finish material now and is even imitated with various materials for its attractive appearance.

An attraction of fired clay is the variety of appearances that can be obtained by using variations of form with the molded material and finishes, both natural and applied as glazes. Add the variations in terms of arrangements of units and form of the mortar joints, and choices are wide-ranging.

2.5   CONCRETE MASONRY UNITS

Masonry units of precast concrete were used to develop some of the earliest applications of concrete to building construction in the nineteenth century. As used today, concrete units (called CMUs) are mostly of forms complying to industry standards and controlled, modular dimensions. A great variety of units is produced, with many variations of forms and finishes for construction exposed to view. Still, the workhorse units are the standard ones of simple form as shown in Fig. 2.4. These are of two basic types, used to produce two types of structural masonry, as follows:

Unreinforced Concrete Block Masonry. This is usually of the single-wythe form shown in Fig. 2.4a. The faces of blocks as well as cross parts are usually quite thick, although thinner units of lightweight concrete are also produced for less serious structural uses. While it is possible to place vertical reinforcement and grout in cavities, this is not the form of block used generally for reinforced construction. Structural integrity of the construction derives basically from unit strength and quality of the mortar. Staggered vertical joints are used to increase the bonding of units.

Reinforced Concrete Block Masonry. This is usually produced with the type of unit shown in Fig. 2.4b. This unit has relatively large individual cavities, so that filled vertical cavities become small reinforced concrete columns. Horizontal reinforcing is placed in courses with the modified blocks shown in Fig. 2.4c or d. The block shown in Fig. 2.4d is also used to form lintels over openings.

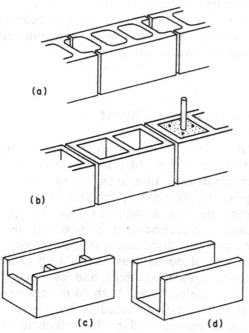

FIGURE 2.4. Standard forms of concrete masonry units.

Concrete units have been produced for many years, which use both ordinary concrete (with sand and gravel) and lightweight concrete with various synthetic aggregates. The lighter units were originally used mostly only for nonstructural applications, but are now also used for reinforced construction. Concrete in general is presently having an expansion of development of new materials, so it is likely that many new types of CMUs will emerge, using fiber reinforcement, superstrength concretes, water-resistive concretes, and other variations on the old standard materials.

As stated previously, the majority of structural masonry is now produced with CMUs, and most of the structural design examples in this book use concrete units. Chapter 5 presents a general discussion of the use of CMUs and the various types of structures produced with them.

2.6 MISCELLANEOUS MASONRY UNITS

Masonry has been produced with a considerable range of units, including blocks of ice, lumps of coal, old bottles, and sea shells. Anything that is reasonably strong, durable, and inert in interaction with mortar can be used. Where building codes must be satisfied, however, it may be hard to make a case for structural masonry of anything other than bricks or concrete blocks. The design examples in this book use only these basic units, but some other forms of masonry are discussed in Chapter 7.

The term *masonry* is capable of extension to a wide range of materials and forms of construction, well beyond the traditional view of its range. With the fairly recent development of reinforced construction and various uses of composites of different materials, new categories continue to emerge. Nonmasonry materials, for example, can be used to form a cast concrete structure like that formed inside concrete blocks in reinforced construction, and the resulting construction is somewhat hard to classify.

2.7 MORTAR

Mortar is usually composed of water, cement, and sand, with some other materials added to make it stickier (so that it adheres to the units during laying up of the masonry), faster setting, and generally more workable during the construction. Building codes establish various requirements for the mortar, including the classification of the mortar and details of its use during construction. The quality of the mortar is obviously important to the structural integrity of the masonry, both as a structural material in its own right and as a bonding agent that holds the units together. While the integrity of the units is dependent primarily on the manufacturer, the quality of the finished mortar work is dependent primarily on the skill of the mason who lays up the units.

There are several classes of mortar established by codes, with the higher grades being required for uses involving major structural functions (bearing walls, shear walls, etc.). Specifications for the materials and required properties determined by tests are spelled out in detail. Still—the major ingredient in producing good mortar is always the skill of the mason. This is a dependency that grows increasingly critical as the general level of craft in construction erodes with the passage of time.

Table 2.1 lists the basic classifications of mortar used by the industry and typically given in building codes. These are derived from a standard, ASTM C270, published by the American Institute for Testing and Material. As specified for construction, a basic property established for structural applications is the specified design strength, designated as f'_m, in units of lb/in.^{2}. This is actually the strength of the *masonry*, which includes considerations for the units and the mortar, and must be verified by tests on assembled

samples of the construction. However, for typical construction, the value is usually assumed, based on the combination of mortar and unit strengths.

2.8 REINFORCEMENT

In the broad sense, reinforcement means anything that is added to help. Structural reinforcement thus includes the use of pilasters, buttresses, tapered form, and other devices, as well as the usual added steel reinforcement. Reinforcement may be generally dispersed or may be provided at critical points, such as at wall ends, tops, edges of openings, and locations of concentrated loads. Both form variation and steel rods are used in both unreinforced masonry and in what is technically referred to as reinforced masonry.

Steel reinforcement is typically used in two forms. One is that made from heavy wire and incorporated in the horizontal mortar joints, as shown in Fig. 2.1. This form is used mostly in what is described as unreinforced construction. The other form uses ordinary, deformed steel bars, the same as those for reinforced concrete

Table 2.1 TYPES OF MORTAR^a

Designation	Description and Use	Average Compressive Strength, 28 days (psi)	(MPa)	Parts by Volume Portland Cement	Masonry Cement	Hydrated Lime
M	High strength. For high stress and all below-grade work.	2500	17.2	1	1	—
				1	—	¼
S	Medium high strength.	1800	12.4	½	1	—
				1	—	¼ to ½
N	Medium strength. For exposed work; minimum required.	750	5.17	—	1	—
				1	—	½ to 1¼
O	Medium low strength. For interior, non-load-bearing work.	350	2.41	—	1	—
				1	—	1¼ to 1½
K	Low strength. For interior, non-load-bearing work.	75	0.52	1	—	2½ to 4

^{a} As defined by ASTM C270.

Table 2.2 PROPERTIES OF STANDARD DEFORMED STEEL REINFORCING BARS

Size	Nominal Diameter		Nominal Area		Nominal Perimeter		Weight	
	(in.)	(mm)	(in.²)	(mm²)	(in.)	(mm)	(lb/ft)	(kg/m)
3	0.375	9.52	0.11	71	1.178	29.92	0.376	0.560
4	0.500	12.70	0.20	129	1.571	39.90	0.668	0.994
5	0.625	15.88	0.31	200	1.963	49.86	1.043	1.552
6	0.750	19.05	0.44	284	2.356	59.84	1.502	2.235
7	0.875	22.22	0.60	387	2.749	69.82	2.044	3.042
8	1.000	25.40	0.79	510	3.142	79.81	2.670	3.973
9	1.128	28.65	1.00	645	3.544	90.02	3.400	5.060
10	1.270	32.26	1.27	819	3.990	101.35	4.303	6.404
11	1.410	35.81	1.56	1006	4.430	112.52	5.313	7.907
14	1.693	43.00	2.25	1452	5.320	135.13	7.650	11.380
18	2.257	57.33	4.00	2581	7.090	180.09	13.600	20.240

construction. Properties of standard deformed bars are given in Table 2.2.

Use of other forms of reinforcement are described in the following chapters.

2.9 LINTELS

A *lintel* is a beam over an opening in a masonry wall. (It is called a *header* when it occurs in framed construction.) For a structural masonry wall the loading on a lintel is usually assumed to be developed only by the weight of the masonry occurring in a 45° isosceles triangle, as shown in Fig. 2.5a. It is assumed that the arching or corbeling action of the wall will carry the remaining wall above the opening, as shown in Fig. 2.5b. There are, however,

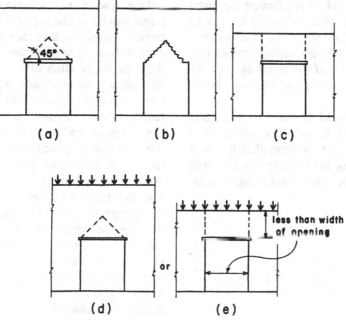

FIGURE 2.5. Considerations for lintels.

many situations that can occur to modify this assumption.

If the wall height above the opening is short with respect to the width of the opening (Fig. 2.5c), it is best to design for the full weight of the wall. This may also be advisable when a supported load is carried by the wall, although the location of the loading and the height of the wall should be considered. If the loading is a considerable distance above the opening, and the opening is narrow, the lintel can probably be designed for only the usual triangular loading of the wall weight, as shown in Fig. 2.5d. However, if the loading is a short distance above the lintel (Fig. 2.5e), the lintel should be designed for the full applied loading; although the wall weight may still be considered in the usual triangular form.

Lintels may be formed in a variety of ways. In times past lintels were made of large blocks of cut stone. In some situations today lintels are formed of reinforced concrete, either precast or formed and poured as the wall is built. With reinforced masonry construction, lintels are commonly formed as reinforced masonry beams, created with U-shaped blocks as shown in Fig. 2.6a when the construction is with hollow concrete units. For heavier loadings, the size of the lintel may be increased by thickening the wall, as shown in Fig. 2.6b.

For unreinforced masonry walls a lintel commonly used is one consisting of a rolled steel section. A form that fits well with a two-wythe brick wall is the inverted tee shown in Fig. 2.6c. Single angles, dou-

ble angles, and various built-up sections may also be used when the loading or the construction details are different.

2.10 ACCESSORIES

In addition to the basic elements of units and mortar, and the usual forms of reinforcement, many special devices are used with masonry construction. Some of the ordinary tasks that these facilitate are the attachment of finish materials, the development of control joints, weather sealing of joints, and achieving of insulation. Use of many of these elements is described in the chapters that follow. Some general considerations for these are described in the following discussions.

Attachment

Attachment of elements of the construction to masonry is somewhat similar to that required with concrete. Where the nature and exact location of attached items can be predicted, it is usually best to provide some built-in device, such as an anchor bolt, threaded sleeve, and so on. Adjustment of such attachments must be considered, as precision of the construction is limited. Attachment can also be effected with drilled-in anchors or adhesives. These tend to be less constrained by the problem of precise location, although the exact nature of the masonry at the point of attachment may be a concern. This is largely a matter of visualization of the complete building construction and of the general problem of integrating the structure into the whole building. It is simply somewhat more critical with masonry structures, as the simple use of nails, screws, and welding is not possible in the direct way that it is with structures of wood and steel.

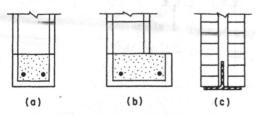

(a)   (b)   (c)

FIGURE 2.6. Forms for lintels.

Control Joints

Shrinkage of mortar, temperature variation, and movements due to seismic actions or settlement of foundations are all sources of concern for cracking failures in masonry. Stress concentrations and cracking can be controlled to some extent by reinforcement. However, it is also common to provide some control joints (literally, preestablished cracks) to alleviate these effects. Planning and detailing of control joints is a complex problem and must be studied carefully as a structural and architectural design issue. Code requirements, industry recommendations, and common construction practices on a local basis will provide guides for this work.

Weather Seals

Airtight, watertight joints are generally required for a building's enclosing construction. Continuous masonry construction may be generally impermeable, but discontinuities created by control joints or connection to other materials require special attention. Masonry generally forms a rigid, nondeforming element, and other materials may shrink, deform structurally, expand thermally at different rates, or otherwise move with respect to the masonry. Weather seals must be achieved without interfering with other functions of joints, such as load transfers, achieving thermal breaks, maintaining fire resistance, and so on. Also, because weather sealing is often done on the outside of joints, appearance may be a major concern.

Insulation

Thermal insulation is most critical in cold climates, and therefore relates to other concerns for cold conditions and to the types of construction generally favored in such regions. Cold climates have the greatest indoor/outdoor temperature differences, and the radiant effects of cold exterior walls on building occupants are critical to comfort. Insulation can be applied as surfacing, incorporated into furred-out spaces, or, in some cases, incorporated into cavities or voids in the masonry, if the masonry construction is not of solid form. Developing of thermal insulation becomes considerably more constrained if the masonry construction is intended to be exposed to view.

3

TYPES OF
MASONRY CONSTRUCTION

Many forms of masonry construction can be used for buildings. This chapter describes the general forms used most commonly and discusses the problems of designing them.

3.1 HISTORICAL FORMS OF CONSTRUCTION

Many forms of masonry construction still in use have long roots to the past. The simple rock piles shown in Fig. 2.3 could have been, and probably were, produced by very primitive people. Basic elements of brick construction are essentially similar today to constructions found in ancient ruins.

Many processes and some materials, however, are quite different. Mortar used today is far superior to that used even in the recent past. Bricks can be produced with greater uniformity with present industrial processes. Some uses, however, are firmly relegated to the past. Huge

structures of cut stone, such as the Egyptian pyramids or Gothic cathedrals, would be virtually impossible to reproduce today. Emergence of new materials, loss of crafts, and spiraling cost of labor have conspired to eliminate many building processes from the available technology.

Study of construction methods, materials, and details used in the past may have some cultural value, but yields very little useful information for development of new construction. The best of the past that is still applicable is preserved in the current technology, but present work that appears historic is mostly derivative and imitative and not really true to the craft traditions. (See Fig. 3.1.)

3.2 CONTEMPORARY CONSTRUCTION

Masonry remains a popular form of construction for many reasons not related to its structural character. Its resistance to

FIGURE 3.1. Forms of construction deeply rooted in the past: adobe and stucco.

fire, rot, and wear, the solid sense it yields due to its rigid nature, and the general character of permanence it implies all make it appealing to the public as quality construction. These attractive attributes are most fully realized when the construction is exposed to view. When the masonry performs strictly utilitarian structural tasks and is out of site—under applied finish materials—it has to compete with other basic structural materials and systems.

The utilitarian masonry structure is almost always of concrete blocks (CMUs),

whether exposed to view or covered up. Other forms, using bricks or stone, are used mostly only when they are to be viewed, and their additional expense can thus be justified.

Just about any form of masonry—ancient or otherwise—can be produced if it can be paid for. If economy must be achieved, the choices are drastically limited. Even if appearance is of major concern, the cost of real brick or stone construction may not be affordable, and a veneered construction must be settled for.

3.3 NONSTRUCTURAL MASONRY

Masonry materials can be used for a variety of functions in building construction. Units of fired clay, precast concrete, cut stone, or fieldstone can be used to form floor surfaces, wall finishes, or non-load-bearing walls (partitions or curtain walls). Indeed, most walls that appear to be made of brick, cut stone, or fieldstone in present-day construction are likely to be of veneered construction, with the surface of masonry units attached to some backup structure.

Although structural utilization of the masonry in these situations may be minor or nonexistent, it is still necessary to develop the construction with attention to many of the concerns given to the production of structural masonry for bearing walls, shear walls, and so on. The quality of the masonry units and the quality of the workmanship of the mortar joints are often just as critical for these uses, although structural behavior or safety may not be at issue. This also extends to concerns for shrinkage, thermal change, stress concentrations at discontinuities, and other aspects of the general behavior of the materials. From an appearance point of view, cracking is just as objectionable in nonstructural masonry facing or paving as it is in a masonry bearing wall.

Masonry veneer facings and non-load-bearing partitions must be provided with control joints and various forms of anchorage and support. Since nonstructural masonry is used extensively, there are many situations for these concerns and a large inventory of recommended construction details. Since these fall mostly in the category of general building construction issues, they are not treated generally in this book.

Masonry units used for nonstructural applications may be of the same structural character as those used for construction of serious masonry structures. Some of the strongest bricks are those that are high-fired to produce great hardness and color intensity for use in veneer construction. However, it is also possible to use some lower structural grades of material, or even some materials that are not usable for structural masonry. Gypsum tile, lightweight concrete blocks, and some very soft bricks may be limited to use in situations not involving structural demands, and probably also not involving exposure to weather or other hazardous situations.

A problem with some nonstructural masonry construction is that of unintended structural response. If the masonry is indeed real masonry (with real masonry units and joints of mortar or grout), it will usually have considerable stiffness. Thus, just as with a plaster surface on a light wood-framed wall structure, it may tend to absorb load due to the relative flexibility of the supporting structure. Much of the cracking of nonstructural masonry (and plaster) is due to this phenomenon. The whole construction must be carefully studied for potential problems of this kind. Flexible attachments, control joints, and possibly some reinforcement can be used to alleviate many of these situations.

A particular problem of the type just mentioned is that of the unintended bracing effect of nonstructural partitions during seismic activity. Actually, there are probably many buildings that have structures that are not adequate for significant lateral loads but are being effectively braced by supposedly nonstructural walls. However, in some cases the general response of the building may be significantly altered by the stiffening effect of rigid partitions.

3.4 STRUCTURAL MASONRY

Many structures of unreinforced masonry have endured for centuries, and this form of construction is still widely used. Although it is generally held in low regard in regions that have frequent earthquakes, it is still approved by most building codes for use within code-defined limits. With good design and good-quality construction, it is possible to have structures that are more than adequate by present standards.

If masonry is essentially unreinforced, the character and structural integrity of the construction are highly dependent on the details and the quality of the masonry work. Strength and form of units, arrangements of units, general quality of the mortar, and the form and details of the general construction are all important. Thus the degree of attention paid to design, to writing the specifications, to detailing the construction, and to careful inspection during the work must be adequate to ensure good finished construction.

There are a limited number of structural applications for unreinforced masonry, and structural computations for design of common elements are in general quite simple. We hesitate to show examples of such work, since the data and general procedures vary considerably from one region to another, depending on local materials and construction practices as well as variations in building code requirements. In many instances, forms of construction not subject to satisfactory structural investigation are tolerated simply because they have been used with success for many years on a local basis—a hard case to argue against. Nevertheless, there are some general principles and typical situations that produce some common problems and procedures. The following discussion deals with some major concerns of design of the unreinforced masonry structure.

Minimal Construction

As in other types of construction, there is a minimal form of construction that results from the satisfaction of various general requirements. Industry standards result in some standardization and classification of products, which is usually reflected in building code designations. Structural usage is usually tied to specified minimum grades of units, mortar, construction practices, and in some cases to need for reinforcement or other enhancement. This usually results in a basic minimum form of construction that is adequate for many ordinary functions, which is usually the intent of the codes. Thus there are many instances in which buildings of a minor nature are built without benefit of structural computations, being simply produced in response to code-specified minimum requirements.

Design Strength of Masonry

As with concrete, the basic strength of the masonry is measured as its resistive compressive strength. This is established in the form of the *specified compressive strength*, designated f'_m. The value for f'_m is usually taken from code specifications, based on the strength of the units and the class of the mortar.

Allowable Stresses

Allowable stresses are directly specified for some cases (such as tension and shear) or are determined by code formulas that usually include the variable value of f'_m. There are usually two values for any situation: that to be used when special inspection of the work is provided (as specified by the code), and that to be used when it is not. For minor construction projects it is usually desirable to avoid the need for the special inspection. For construction with hollow units that are not fully grouted, stress computations are based on the net cross section of the hollow construction.

Avoiding Tension

While codes ordinarily permit some low stress values for flexural tension, many designers prefer to avoid tension in unreinforced masonry. An old engineering definition of mortar is "the material used to keep masonry units *apart*," reflecting a lack of faith in the bonding action of mortar.

Reinforcement or Enhancement

The strength of a masonry structure can be improved by various means, including the insertion of steel reinforcing rods. Vertical rods are sometimes used with construction that is essentially classified as unreinforced, usually to enhance bending resistance or to absorb localized stress conditions. Horizontal wire-type reinforcing is commonly used to reduce stress effects due to shrinkage and thermal change.

Another type of structural reinforcement is achieved through the use of form variation, examples of which are shown in Fig. 3.2. Turning a corner at the end of a wall (Fig. 3.2a) adds stability and strength to the discontinuous edge of the structure. An enlargement in the form of a pilaster column (Fig. 3.2b) can also be used to improve the end of a wall or to add bracing or concentrated strength at some intermediate point along the wall. Heavy concentrated loads are ordinarily accommodated by using pilaster columns when wall thickness is otherwise minimal. Curving a wall in plan (Fig. 3.2c) is another means of improving the stability of the wall.

Building corners and openings for doors or windows are other locations where enhancement is often required. Figure 3.2d shows the use of an enlargement of the wall around the perimeter of a door

opening. If the top of the opening is of arched form, the enlarged edge may continue as an arch to span the opening, as shown in Fig. 3.2e, or a change may be made to a stronger material to effect the edge and arch enhancement, as shown in Fig. 3.2f. Building corners in historic buildings were often strengthened by using large cut stones to form the corner, as shown in Fig. 3.2g.

While form variations or changes of the masonry units can still be used to achieve spans over openings, a lintel is usually used. Flat-spanning lintels, consisting usually of cut stone, were used in the past. The spanning capability of the stone was drastically limited by its low tensile strength, as discovered by early temple builders, resulting in the close spacing of columns, an architectural feature of Egyptian, Greek, and Roman temples. As discussed in Sec. 2.9, lintels are now more often achieved as reinforced masonry or concrete or by insertion of steel elements, as shown in Fig. 3.2h.

Figure 3.3 shows various details of masonry structures with details reminiscent of ancient construction. Reinforcement of corners and openings is achieved with units apparently of cut stone, as would be done in earlier times. These buildings are mostly from the early twentieth century, however, and the "stone" is actually precast concrete in all the photos.

3.5 UNREINFORCED CONSTRUCTION

In contemporary construction, so-called (by the building code) unreinforced masonry is seldom so—if reinforcement by any means described in the preceding section is accepted as being real. Even steel reinforcement is used in much of the construction classified as unreinforced in engineering terms. Thus, it must be borne in

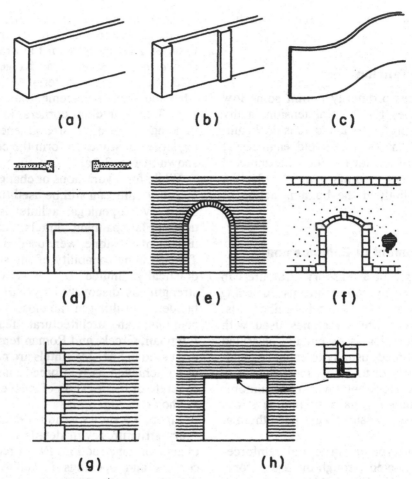

FIGURE 3.2. Forms of reinforcement in masonry construction.

mind that the building code classification of reinforced construction is limited to a particular form of construction.

A major point of distinction that should be recognized, however, is that when the construction is essentially not reinforced, its basic structural character derives primarily from the masonry alone. Thus the integrity of the units, strength and adhesion of the mortar, and absence of cracking or other flaws in the work become more significant. This requires some heightened attention to careful specification of materials and construction processes and to continuous supervision of the work during construction. No matter how good the materials and how carefully done the construction details may be, good masonry work is still largely in the hands of the masons, and masonry construction is still linked to craft-intensive processes to a degree not largely present in most contemporary work of building construction.

Structural design of masonry elements is really quite simple in most cases. The greatest attention and predominance of publications deals with the quality of materials and the care in the construction processes. Structural design that does not extend to dealing with these aspects is not likely to be effective.

FIGURE 3.3. Corners in brick walls achieved with stone (actually precast concrete) units.

3.6   REINFORCED CONSTRUCTION

As we use it here, the term *reinforced masonry* designates a type of masonry construction specifically classified by building code definitions. Essential to this definition are the assumptions that the steel reinforcement is designed to carry forces and the masonry does not develop tensile stresses. This makes the design basically analogous to that for reinforced concrete, and indeed the present data and design procedures used for reinforced masonry are in general similar to those used for concrete structures. Until recently, the general methods of the working stress design procedure were used for masonry. However, recent industry standards and some building codes (including the 1988 edition of the *UBC*) have promoted the use of strength methods for investigation and design. As with concrete, the use of strength methods is quite complex and abstract. We have therefore chosen to use the simpler methods of the working stress design procedures for most of the example computations to permit a briefer treatment; our interest being more in demonstrating the problem and the basic concerns for investigation rather than particular means for investigation.

Reinforced Brick Masonry

Reinforced brick masonry typically consists of the type of construction shown in Fig. 2.1. The wall shown in the illustration consists of two wythes of bricks with a cavity space between them. The cavity space is filled completely with grout so that the construction qualifies firstly as *grouted masonry*, for which various requirements are stipulated by the code. One requirement is for the bonding of the wythes, which can be accomplished with the masonry units, but is most often done by using the steel wire joint reinforcement shown in the illustration in Fig. 2.1.

Added to this basic construction are the vertical and horizontal reinforcing rods in the grouted cavity space, making the resulting construction qualify as *reinforced grouted masonry*.

General requirements and design procedures for the reinforced brick masonry wall are similar to those for concrete walls. There are stipulations for minimum reinforcement and provisions for stress limits for the various structural actions of walls in vertical compression, bending, and shear wall functions. Structural investigation is essentially similar to that for the hollow unit masonry wall, which is discussed in the next section.

Despite the presence of the reinforcement, the type of construction shown in Fig. 2.1 is still essentially a masonry structure, highly dependent on the quality and structural integrity of the masonry itself—particularly the skill and care exercised in laying up the units and handling of the construction process in general. The grouted, reinforced cavity structure is in itself often considered to be a third wythe, constituted as a very thin reinforced concrete wall panel. The enhancement of the construction represented by the cavity wythe is considerable, but the major bulk of the construction is still basically just solid brick masonry.

Reinforced Hollow Unit Masonry

Reinforced hollow unit masonry most often consists of single-wythe walls formed as shown in Fig. 3.4. Cavities are vertically aligned so that small reinforced concrete columns can be formed within them. At some interval, horizontal courses are also used to form reinforced concrete members. The intersecting vertical and horizontal concrete members thus constitute a rigid frame bent inside the wall. This reinforced concrete frame is the major structural component of the construction. Besides providing forming, the concrete

FIGURE 3.4. Common form of reinforced masonry with CMUs.

blocks serve to brace the frame, provide protection for the reinforcement, and interact in composite action with the rigid frame. Nevertheless, the structural character of the construction derives largely from the concrete frame created in the void spaces in the wall.

The code requires that reinforcement be a maximum of 48 in. on center; thus the maximum spacing of the concrete members inside the wall is 48 in., both vertically and horizontally. With 16-in.-long blocks this means that every sixth vertical void space is grouted. With blocks having a net section of approximately 50% (half solid, half void), this means that the minimum construction is an average of approximately 60% solid. For computations based on the net cross section of the wall, the wall may thus usually be considered to be a minimum of 60% solid.

If all void spaces are grouted, the construction is fully solid. This is usually required for structures such as retaining walls and basement walls, but may also be done simply to increase the wall section for the reduction of stress levels. Finally, if reinforcing is placed in all of the vertical voids (instead of every sixth one), the contained reinforced concrete structure is considerably increased and both vertical bearing capacity and lateral bending capacity are significantly increased. Heavily loaded shear walls are developed in this manner.

For shear wall actions, there are two conditions defined by the code. The first case involves a wall with minimum reinforcing, in which the shear is assumed to be taken by the masonry. The second case is one in which the reinforcing is designed to take all of the shear. Allowable stresses are given for both cases, even though the reinforcement must be designed for the full shear force in the second case.

Design of typical elements of reinforced masonry is discussed in the next two chapters.

Using the analogy of the fully grouted and reinforced hollow unit masonry, it is possible to visualize a construction using just about anything to form a wall that derives its basic structure from being filled with concrete and steel to achieve an internal rigid frame. In truth, many such special constructions exist, using foam plastic and other inert forming units to hold the cast concrete. The forms may thus be ignored structurally, and the structure recognized simply as a concrete frame. A variation on this is to use any hollow formed units of precast concrete and fill them up with sitecast concrete and reinforcement to achieve a structure. This is not exactly masonry construction, but it owes a real kinship to the concrete-filled construction with hollow masonry units.

3.7 VENEERED CONSTRUCTION

A veneer is a finish coating or layer of material that achieves the appearance of solid, natural material. The veneer may be a paper-thin layer of walnut over a structural softwood plywood or a single wythe of bricks over some structural, supporting backup. Masonry veneers often strive to achieve the appearance of traditional, structural masonry. The vast majority of new construction that appears to be structural masonry of brick or stone is actually veneered construction. Purists disavow veneered construction, but it is nevertheless widely used.

The masonry "look" can be achieved in a variety of ways, including coverings with formed sheets of plastic, thin adhered tiles, and fiber-reinforced concrete elements formed and colored to simulate

bricks or stone. "Cast stone" that simulates polished granite or marble is used in large, thin panels as a relatively lightweight infill element with curtain wall systems. The latter constructions are essentially imitations of masonry. Without any philosophical judgment involved, the work in this book is generally limited to uses of "real" masonry. The remainder of this discussion therefore deals with veneers that use actual bricks or stone set in real mortar.

Veneers of brick as thick as 3 to 4 in. are nevertheless too thin to be freestanding if a full story in height. As a veneer, the single wythe of brick needs lateral support, and must be attached in some way to a supporting structure. The support may be provided by virtually any structure,

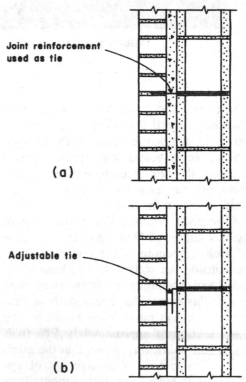

FIGURE 3.5. Brick exterior: (a) structurally joined to CMU backup for composite construction; (b) as a nonstructural veneer, tied to the CMU backup for lateral support.

and indeed does range from actual masonry (usually CMUs) to a light wood frame. With the masonry backup it is possible to consider a continuous bonding of the veneer and the backup in a monolithic construction. (See Fig. 3.5a.) However, it is more common to use a veneer that is separated by an airspace and tied across the space to the backup (Fig. 3.5b). This airspace adds some insulative quality to the construction, although it must be vented at the top and drained at the bottom to avoid collection of water condensate.

Most veneers are not structural, but are sometimes associated with a structural masonry backup, requiring some considerations in the design and detailing of the supporting masonry for proper relation to the veneer. Since this book deals essentially with structural masonry, the discussions of veneers in the remainder of this book are limited to considerations involving their relations to supporting masonry structures.

3.8 STRUCTURAL DESIGN STANDARDS

Masonry structures are most often used for relatively simple tasks, and their investigation and design are correspondingly not very complicated. Unreinforced masonry is used mostly for simple compression members (bearing walls, piers, pedestals, columns) and develops mostly only simple direct compression or some combined bending and compression.

Reinforced masonry has a character that emulates reinforced concrete, and analytical methods used are generally borrowed from those used for reinforced concrete structures. The working stress methods are still widely used, although most codes now provide for some design with strength and factored load methods.

Investigations and design procedures illustrated in this book are limited to relatively simple and common structural elements. For this work, the following basic references have been used:

Uniform Building Code, 1988 edition (Ref. 1).

Building Code Requirements for Masonry Structures, ACI 530-88 (Ref. 4).

Masonry Design Manual, 4th edition (Ref. 7).

Reinforced Masonry Design, 3rd edition (Ref. 8).

Plus, as cited in the text, miscellaneous publications of the

Brick Institute of America and
National Concrete Masonry
  Association.

4

BRICK MASONRY

Bricks were developed in ancient times as blocks of dried mud. If simply allowed to dry into a hardened form, they were called *sun-dried* or *adobe* bricks. When the general process of firing clay materials to produce ceramics was developed, bricks were produced as *fired* or *burned* bricks. This is still the basic way that bricks are produced, using soil materials that are predominantly clay. Adobe construction sees only limited use today, and is described in Sec. 7.1. The discussion in this chapter deals with construction developed with commercially produced, fired clay bricks. Inexpensive ''bricks'' can also be produced of precast concrete, but the real thing is still considered to be the fired clay brick.

4.1 TYPES OF BRICKS

Many bricks are produced for usages that are not structural. These are called *facing* bricks, and standards for them deal mostly with control of appearance quality. Bricks produced for structural use are called *building* bricks, and are generally made to conform to the classifications of ASTM standard C216. Table 4.1 lists the three grades of building brick described in the ASTM standard and the general conditions for their use.

Bricks are ordinarily produced in a flat, elongated form, as shown in Fig. 4.1, intended for placement in wall construction in the position shown in the figure. For this use, the sides of the brick are referred to by the terms indicated. The *face* of the brick refers to its normal position in a wall, with the side that is viewed on the face of the wall. Obviously, when the wall surface is exposed to view, the brick face becomes of primary concern, both for its dimensions and the visible color and texture of the brick material.

There is no such thing as a standard brick with regard to dimensions, although some very common dimensions are used. Table 4.2 lists some sizes that are used for

Table 4.1   CLASSIFICATION OF GRADE OF BUILDING BRICKS

Classification Designation (ASTM C216)	General Exposure Conditions	Basic Usage Conditions
SW	Severe weathering	Below grade, or any exposed condition in severe weather regions
MW	Moderate weathering	Above-grade exposures, mild climate regions
NW	Negligible weathering	Indoor or weather-sheltered conditions

various purposes. Regional differences and specific products of manufacturers present considerable variety in the range of actual materials available.

The modular brick described in Table 4.2 is developed for use in construction that generally conforms to a three-dimensional 8-in. layout module, which also frequently relates to use with concrete blocks of standard size (nominal 8 in. by 8 in. by 16 in. as the standard structural unit).

The roman brick emulates the large, thin bricks used by Roman builders. One reason for making them thin was simply that they could be dried and used faster (to make a fortification before the enemy troops arrived for battle).

The jumbo brick was developed for use in single-wythe, nominal 6-in.-thick walls, mostly for small, low-rise construction. Because of its large size, it is usually made with voids, in the general manner of clay tile or concrete blocks.

4.2   ARRANGEMENT OF UNITS IN BRICK WALLS

As indicated in Fig. 4.1, the ordinary position for a brick in wall construction is that in which it lies on its flat side, with the long, thin side (face) exposed. When so positioned, it is referred to as a *stretcher*. For various purposes, however, bricks may sometimes be placed in other positions, as shown in Fig. 4.2.

Standard-shape bricks are normally used in walls of at least two wythes. For stability of the construction, it is necessary to tie the wythes together (generally called *bonding*). If this is achieved only through use of the masonry units, the common element used as a tie in brick construction is a brick turned to expose its end in the wall face, as shown in Fig. 4.2*a*. If the brick is laid flat-side down, it is called a *header*. If it is laid face-side down, as shown in Fig. 4.2*b*, it is called a *rowlock*.

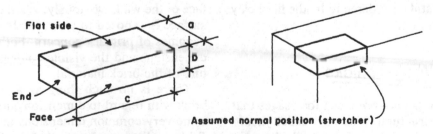

FIGURE 4.1. Reference terms for bricks.

Table 4.2 COMMON SIZES OF BRICKS

| | Dimensions (in.)[a] | | | |
Description	Thickness *a*	Height *b*	Length *c*	Usage
Standard	$3\frac{3}{4}$	$2\frac{1}{4}$	8	General
Modular	$3\frac{5}{8}$	$2\frac{1}{4}$	$7\frac{5}{8}$	Construction with 3D 8-in. modular planning
Roman	$3\frac{5}{8}$	$1\frac{5}{8}$	$11\frac{5}{8}$	2-in. vertical, 12-in. horizontal module on wall face
Jumbo	$5\frac{5}{8}$	$3\frac{5}{8}$	$11\frac{5}{8}$	For single-wythe, nominal 6-in.-thick walls

[a] See Fig. 4.1.

The other basic position commonly used in traditional brick masonry is the *soldier* position, shown in Fig. 4.2c. This has no particular function, being essentially an architectural device, sometimes used to accent the top of a wall or the top or bottom of an opening.

Based on the most common positions of stretchers and headers, some classic forms of wall face patterns were developed by various ancient builders. These are shown in Fig. 4.3. Structural masonry with bricks is now commonly tied with metal elements (see the joint reinforcement in Fig. 3.1), but patterns often still reflect the old styles.

A special pattern is the stack bond, in which units are simply arranged in neat rows, horizontally and vertically. This is obviously not a sound pattern for structural masonry, if the bonding of the construction is to be achieved solely by the masonry units and mortar joints. However, if binding of wythes is fully achieved with joint reinforcement, and the construction is fully grouted and reinforced, it can be an acceptable structural form of construction.

From these basic patterns, many variations are possible. Indeed, great variety was achieved by ancient masons, using almost every trick known today. Simple

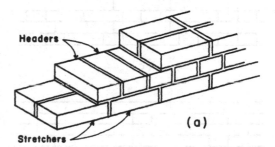

(a)

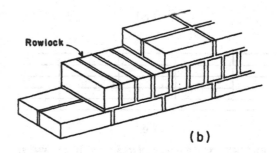

(b)

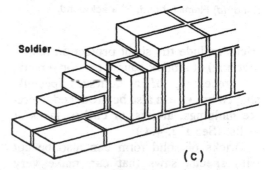

(c)

FIGURE 4.2. Ordinary positions for bricks.

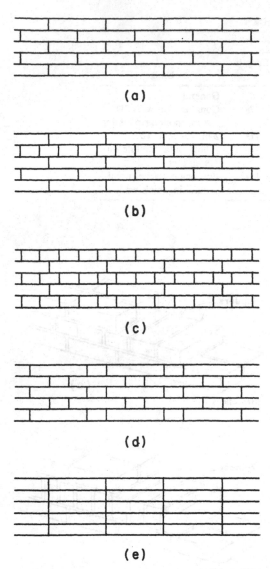

(a)

(b)

(c)

(d)

(e)

FIGURE 4.3. Face patterns for brick walls: (a) running bond; (b) common bond; (c) English bond; (d) Flemish bond; (e) stack bond.

tricks include the slight projection (cantilevering) of a brick course or, conversely, the recessing of a course. Specially shaped bricks can also be produced to create moldings, arches, or curved faces of walls. (See Fig. 4.4.)

Bricks of solid form can also be cut with special saws that can make very clean, accurate cuts. This somewhat frees the dimensions of the finished construction from a tight relationship to the specific brick dimensions. This is generally true for horizontal dimensions. However, it is less feasible to cut bricks to adjust their height, so vertical coursing should be planned to relate to the vertical brick dimension (height in Fig. 4.1) and the selected dimension of mortar joints.

4.3 MORTAR JOINTS

In the past, it was considered to be high craftsmanship for a mason to achieve very thin mortar joints, Today the general brick mortar joint is ⅜ in. thick. The modular dimensions given in Table 4.2 assume this size joint. One reason for the thicker joints today is to accommodate the insertion of joint reinforcement and some other metal accessories used for anchoring of elements to structural walls.

For work in which the wall face is not exposed to view, the face of the mortar joint is simply cut back to be reasonably smooth in alignment with the brick face. Actually, a better structural joint is achieved if the mortar is allowed to ooze out of the joint, ensuring a better full contact with good mortar to the full flat side of the bricks. The smooth-faced wall is preferable if materials are to be attached to the wall—for example, furring strips or solid blocks of insulation.

If exposed to weather, it is usually more desirable to have a *tooled joint*. This is generally achieved by first striking the joint face flush with the wall surface as the bricks are laid. Then when the mortar is stiffened somewhat, but still moldable, a metal tool is used to shape the joint face. Various profiles for the tooled joints are shown in Fig. 4.5. Tooling serves the practical purpose of slightly compacting the mortar at the face, presenting a denser, more weather-resistive surface. The profile should be one that drains well, with those labeled from (*b*) to (*d*) in Fig. 4.5 being preferred for severe weather exposure.

The mortar joints represent a major element in the appearance of brick construction. This brings heightened concern for both the color of the mortar and the form of the joint profile. Mortar materials are generally available in color tones and are selected for a blend with the brick units. Inset profiles of the tooled joints create shadow lines that emphasize joints. These

FIGURE 4.4. Variations of form in brick masonry.

FIGURE 4.4. (*Continued*).

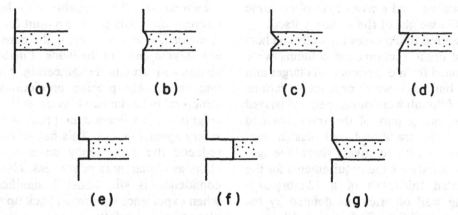

FIGURE 4.5. Profiles of mortar joints: (a) flush; (b) concave; (c) vee; (d) weathered; (e) raked; (f) stripped; (g) struck.

can be heightened visually by use of some special joints, such as those labeled (e) through (g) in Fig. 4.5. These are not recommended for severe weather exposures, however.

4.4 TYPICAL ELEMENTS OF BRICK CONSTRUCTION

For structural purposes, brick is used primarily to produce walls, piers, columns, and pedestals. In masonry terminology, *piers* are segments of walls, short in plan length, while *pedestals* are very short columns. There is thus a sort of geometric order of progression for a singular element that is basically a compression member oriented for vertical loading. (See Fig. 4.6.)

Structural masonry, in the form of walls or columns, is often used for support of horizontal-spanning structures and possibly for support of other walls or columns above in multistory construction. A major consideration, therefore, is that for resistance to vertical gravity loads that induce compression stress in the walls or columns. In ancient, predominantly very heavy, masonry, a significant amount of

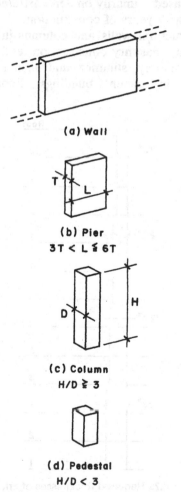

(a) Wall

(b) Pier
3T < L ≦ 6T

(c) Column
H/D ≧ 3

(d) Pedestal
H/D < 3

FIGURE 4.6. Classification of vertical compression members.

the total vertical gravity load often came from the weight of the masonry itself.

Required thicknesses for walls and horizontal cross sections for columns were developed from experience. As larger and taller buildings were produced, various rules of thumb were developed and passed on, becoming part of the traditions and craft. This tradition-based design was dominant until a relatively short time ago. Figure 4.7 shows the requirements for the graduated thickness of a 12-story-high bearing wall of brick, as defined by the Chicago Building Code some 60 years ago. This criteria was largely judgmental and based primarily on demonstrated success with years of construction.

Structural walls and columns in present-day masonry construction tend to be considerably slimmer and lighter than those in ancient buildings. Economic pressures and the availability of better, stronger materials partly account for this. A major factor, however, is the considerable development of the fields of materials testing and structural engineering, permitting reasonably precise predictions for structural behaviors and, as a result, some heightened confidence in predictions of safety against failure. This has somewhat reduced the dependency on experience alone as a guarantee of success, although confidence is still boosted significantly when experience is there to back up speculations about safety.

Actually, in spite of the lack of sophisticated structural analysis and design theories and procedures, ancient builders produced many technological feats of daring with masonry construction that generally cannot be duplicated today, even though some of their achievements are still standing as testimonials to the builders' successes. Structures like the Roman aqueducts and the Gothic cathedrals would never be executed in masonry today, despite our considerable advances in technology of materials and the science of materials behaviors. Structures can be built to look like the ancient masonry ones, but will certainly be actually achieved with steel or reinforced concrete.

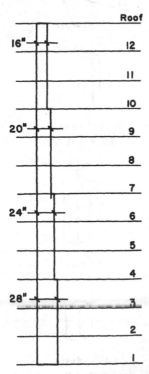

FIGURE 4.7. Required thicknesses of brick walls for a 12-story mercantile building, Chicago Building Code, 1932.

4.5 GENERAL CONCERNS FOR BRICK WALLS

Structural brick walls can be used for many purposes, and basic design concerns derive initially from the specific structural requirements. Compression stress from vertical loading is a predominant condition, deriving—if from nothing else—from the weight of the masonry itself. For simple bearing walls, this may be the only real concern, and a simple axial compression stress investigation may suffice for the structural design. Thus the total, critical compression load is determined, and the simple stress calculation

$$f = \frac{P}{A}$$

is made, in which

  f = average compression stress
  P = total vertical compression force
  A = cross-sectional area of the wall

This calculated stress is compared to an allowable stress for the construction, based on code requirements and the specifications for the masonry materials.

For solid masonry construction, the wall cross section is simply the product of the wall thickness and some defined, design increment of the wall length (typically a 1-ft or 12-in. length in U.S. units). For cavity walls, or walls made with hollow units, the actual net solid cross section must be used.

Additional structural functions for walls may include any of those illustrated in Fig. 4.8. Exterior walls in buildings frequently must serve multiple purposes: bearing walls, spanning walls resisting direct wind pressures, and shear walls. These situations produce multiple concerns for stress combinations, as well as for the logical development of wall form, construction details, attachment of any supported elements, and so on. Structural walls are also themselves supported by other structures, either directly by foundation elements or by other supporting walls or frameworks.

In the past, brick construction was frequently used as a "cheap" infill material to achieve solid masonry with higher-quality surface materials. Now brick has graduated mostly to the higher class itself, and its use must usually be justified by exposure of the expensive construction. Thus, appearance considerations become major concerns, even when serious struc-

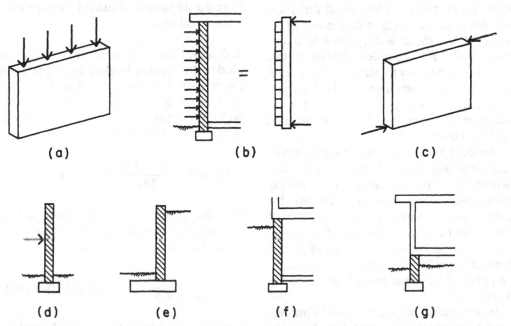

FIGURE 4.8. Structural uses for masonry walls: (a) bearing; (b) spanning for wind; (c) shear bracing; (d) freestanding; (e) cantilever retaining; (f) foundation—basement; (g) foundation—grade.

tural tasks are required. Brick materials, mortar color and joint form, and wall face patterns are typically major architectural design concerns, as are various construction details if exposed to view. Wall ends, supports, and tops, and edges of openings for windows and doors must be sensitively detailed, both for appearance and for the special structural tasks at these locations.

For various reasons, brick walls are mostly used in only a few common ways, especially for structural purposes. The following sections describe some specific uses and some of the additional concerns that derive from them. Usage considerations in a broader context are also discussed in the building design cases in Chapter 10.

4.6 BRICK BEARING WALLS

Brick bearing walls often occur as exterior walls. They thus quite frequently have other structural functions, in addition to that of resisting simple vertical compression. Three common additional functions are those shown in Fig. 4.8b (spanning for wind), 4.8c (shear wall), and 4.8f (basement wall). For complete design it thus becomes necessary to consider these combined actions. Some combined effects will be discussed later in this section, but we will first consider the simple problem of vertical compression.

Readers with less experience in structural investigation should review Appendix A before pursuing the following discussions. The general concerns for compression members are presented there, and the various methods for investigation are discussed. Additionally, the general basis for working-stress investigation of reinforced members is presented in Appendix B.

Investigation of compression in unreinforced construction may be done by using maximum stress limits, such as those in

Table 4.3, which is a reprint of Table 24-H from the 1988 *UBC* (Ref. 1). This limiting stress may be used as long as the wall slenderness does not exceed limits established by the code. When slenderness is excessive, a general formula for allowable compression stress is

$$F_a = 0.20f'_m \left[1 - \left(\frac{h'}{42t} \right)^3 \right]$$

in which

F_a = allowable average compressive stress

h' = effective unbraced height

t = effective thickness

Example 1. A brick wall of solid construction is 9 in. thick and has an unbraced height of 10 ft. The construction has a specified value of $f'_m = 2500$ psi and is unreinforced. The wall sustains a bearing load on its top of 2000 lb/ft. The brick construction weighs approximately 140 lb/ft³. Investigate the condition for average compressive stress.

Solution: The total load on the wall includes the applied loading on the top plus the weight of the masonry. This total gravity load is greatest at the bottom of the wall. We thus determine the total wall weight for a 1-ft (12-in.)-wide strip to be

$$\text{Weight} = \frac{9 \times 12}{144}(10)(140) = 1050 \text{ lb}$$

The total load is then 1050 + 2000 = 3050 lb, and the maximum compressive stress at the bottom of the wall is

$$f_a = \frac{P}{A} = \frac{3050}{9 \times 12} = 28.2 \text{ psi [194 kPa]}$$

Since the allowable-stress formula is based on consideration for the wall slen-

Table 4.3 ALLOWABLE WORKING STRESSES IN UNREINFORCED UNIT MASONRY

MATERIAL	TYPE M Compression1	TYPE S Compression1	TYPE M OR TYPE S MORTAR Shear or Tension in Flexure2 3		Tension in Flexure4		TYPE N Compression1	TYPE N Shear or Tension in Flexure2 3	
1. Special inspection required	No	No	Yes	No	Yes	No	No	Yes	No
2. Solid brick masonry									
4500 plus psi	250	225	20	10	40	20	200	15	7.5
2500–4500 psi	175	160	20	10	40	20	140	15	7.5
1500–2500 psi	125	115	20	10	40	20	100	15	7.5
3. Solid concrete unit masonry									
Grade N	175	160	·12	6	24	12	140	12	6
Grade S	125	115	12	6	24	12	100	12	6
4. Grouted masonry									
4500 plus psi	350	275	25	12.5	50	25			
2500–4500 psi	275	215	25	12.5	50	25			
1500–2500 psi	225	175	25	12.5	50	25			
5. Hollow unit masonry5	170	150	12	6	24	12	140	10	5
6. Cavity wall masonry solid units5									
Grade N or 2500 psi plus	140	130	12	6	30	15	110	10	5
Grade S or 1500–2500 psi	100	90	12	6	30	15	80	10	5
Hollow units5	70	60	12	6	30	15	50	10	5
7. Stone masonry									
Cast stone	400	360	8	4	—	—	320	8	4
Natural stone	140	120	8	4	—	—	100	8	4
8. Unburned clay masonry	30	30	8	4	—	—	—	—	—

[1]Allowable axial or flexural compressive stresses in pounds per square inch gross cross-sectional area (except as noted). The allowable working stresses in bearing directly under concentrated loads may be 50 percent greater than these values.

[2]This value of tension is based on tension across a bed joint, i.e., vertically in the normal masonry work.

[3]No tension allowed in stack bond across head joints.

[4]The values shown here are for tension in masonry in the direction of running bond, i.e., horizontally between supports.

[5]Net area in contact with mortar or net cross-sectional area.

Source: Table 24-H in the *Uniform Building Code*, 1988 ed., reproduced with permission of the publishers, International Conference of Building Officials.

derness, it is technically critical near the wall mid-height (where buckling occurs). However, for a conservative investigation we can compare the calculated stress just determined with that found from the formula for allowable stress, as follows:

$$F_a = 0.20 f'_m \left[1 - \left(\frac{h'}{42t} \right)^3 \right]$$

$$= 0.20(2500) \left[1 - \left(\frac{120}{42 \times 9} \right)^3 \right]$$

$$= 484 \text{ psi}$$

This value is compared with the maximum limit given in Table 4.3. For a minimum Type S mortar, the table yields a maximum value of 160 psi. Although this is lower than that determined from the formula, it is still greater than the computed average stress, so the wall is still safe.

Allowable stresses for design may be modified by various conditions, one of which relates to the type of inspection and testing provided for the construction. The design code used must be carefully studied for these qualifications.

When walls sustain concentrated loads, rather than distributed loads, it is usually necessary to investigate for two stress conditions. The first involves the concentrated bearing stress directly beneath the applied load. The second investigation involves a consideration of the effective portion of the wall that serves to resist the load. The length of wall for the latter situation is usually limited to six times the wall thickness, four times the wall thickness plus the actual bearing width, or the center-to-center spacing of the loads. The following example demonstrates this type of investigation.

Example 2. Assume that the wall in Example 1 sustains a concentrated load of 12,000 lb from the end of a truss. The truss load is transferred to the wall through a steel bearing plate that is 2 in. narrower than the 9-in.-thick wall and 16-in.-long along the wall length. Investigate for bearing and average compression stresses in the wall.

Solution: Using the given data, we first determine the actual values for the bearing and average compression stresses. For the bearing,

$$f_{br} = \frac{P}{A} = \frac{12,000}{7 \times 16} = 107 \text{ psi } [739 \text{ kPa}]$$

For the compression, we assume an effective wall unit that is $6 \times 9 = 54$ in. or $(4 \times 9) + 16 = 52$ in., the latter being critical. Using the unit density given in

Example 1, we find that this portion of the wall weighs

$$(140) \frac{9 \times 52}{144} (10 \text{ ft}) = 4550 \text{ lb}$$

When this is added to the applied load, the total vertical compression at the base of the wall is thus $12,000 + 4550 = 16,550$ lb, and the average compressive stress is

$$f_a = \frac{P}{A} = \frac{16,550}{9 \times 54} = 34.0 \text{ psi } [234 \text{ kPa}]$$

The allowable compression stress for this case is the same as that found in Example 1, 160 psi. The actual stress is thus well below the limit for safety.

For the bearing stress, the code provides two values, one based on "full bearing," and the other on a defined condition where the bearing contact area is a fraction of the full cross section of the masonry. Although the actual bearing area in this case is indeed a fraction of the full wall cross section, the short distance from the edge of the bearing area to the edge of the masonry qualifies this situation as full bearing. We therefore use the code limit for allowable bearing as

$$F_{br} = 0.26f'_m = (0.26)(2500)$$
$$= 650 \text{ psi } [4482 \text{ kPa}]$$

which indicates a safe condition for bearing for the example.

Solid brick masonry is usually the strongest form of unreinforced masonry, and thus is capable of considerable force resistance for situations involving only simple compression and bearing stresses.

Design of unreinforced masonry columns generally follows the same procedures just demonstrated for the bearing wall. From the data of Example 2, it may be demonstrated that a 9-in.-square column could sustain the load, using the allowable stress limit of 160 psi. While this may satisfy code limits, a 10-ft-high column only 9 in. across would appear quite slender. The author does not recommend the use of unreinforced columns in general, except for very low magnitudes of load and columns in the pedestal class, where the column height is less than three times its side dimension.

Compression Plus Bending

From eccentrically placed compression loads or some form of direct bending action, compression members frequently must resist a combination of direct compression plus flexure. The general case for this may be visualized as that occurring with the eccentric compression force, which induces a bending moment equal to the product of the force and its eccentricity from the compression member's centroidal axis. See discussions in Appendix A, Secs. A.5 and A.8.

The condition of net stress caused by a combination of compression and bending can be visualized as an addition of the separate stresses of uniformly distributed compressive stress plus the varying stresses caused by bending, ranging from a maximum tension to a maximum compression across the stressed section. In a member capable of resisting both compression and tension, this will result in some limiting, boundary values of the stress variation on the section. A special case occurs when the section is actually not capable of tension resistance—a conservative assumption for masonry, which has a very low tensile resistance. An analogy can be made with bearing foundations on soil, where the foundation–soil contact face can resist only compression.

The discussion in Appendix A (Sec. A.8) develops some relationships for investigation of compression plus bending occurring with a simple bearing footing, a

condition we will use to represent the tension-weak, unreinforced masonry.

When tension stress is not possible, eccentricities beyond the kern limit will produce a *cracked section,* which is shown as case 4 in Fig. A.7. In this situation some portion of the section becomes unstressed, or cracked, and the compressive stress on the remainder of the section must develop the entire resistance to the force and moment.

The cracked-section stress condition is really not desirable for either soil bearing or masonry. The kern limit, with a zero stress at one edge, should be used in most cases as a design limit. A possible exception may be for stress combinations that include wind or seismic effects, which are extremely short in duration. Masonry is now generally permitted to sustain a very low tensile stress, but it is still better to design without it.

The following example illustrates an application of the combined stress relationships derived in Appendix Sec. A.8.

Example 3. Investigate the wall in Example 1 for a combined loading that includes the gravity loads as given plus a wind pressure of 20 psf on the wall surface.

Solution: The stress due to the vertical gravity load was found in Example 1. That is actually the value of the compressive stress at the base of the wall, since the load includes the entire wall weight. Bending stress will have its greatest value at mid-height of the wall, as the following discussion indicates.

The wall spans vertically a distance of 10 ft (its unbraced height) and sustains a uniformly distributed load on a 1 ft wide strip of 20 lb/ft. Using the formula for maximum moment on a simple beam, we find the maximum bending moment at mid-height in the wall:

$$M = \frac{wL^2}{8} = \frac{(20)(10)^2}{8} = 250 \text{ ft-lb}$$

For the solid rectangular section, 9 in. by 12 in., the section modulus is determined to be

$$S = \frac{bd^2}{6} = \frac{(12)(9)^2}{6} = 162 \text{ in.}^3$$

and the maximum bending stress is found to be

$$f_b = \frac{M}{S} = \frac{250 \times 12}{162} = 18.5 \text{ psi}$$

Adding this stress to the axial compression stress produces

$$f = f_a + f_b = 28.2 \pm 18.5$$

$$= 46.7 \text{ psi in net compression (maximum)}$$

$$= 9.7 \text{ psi in net compression (minimum)}$$

(See case 1 in Fig. A.7.)

This is obviously not a critical stress condition, especially since the load combination including wind permits the allowable stress of 160 psi (see Example 1) to be increased by one third.

As previously described, the wind load produces a simple beam-bending action in the wall with a maximum bending moment at mid-height, as shown in Fig. 4.9a. The stress combination just determined is thus conservative, since it uses the maximum compression at the base of the wall. However, the low-stress condition does not justify a more accurate investigation.

Bending moments in walls can also be induced by other loading conditions. A common situation is that shown in Fig. 4.9b, in which a supported gravity load is placed off the wall centroid (in this case, the center of the wall). This may occur when a wall is continuous past a floor level in multistory construction or when a wall forms a parapet by extending past a

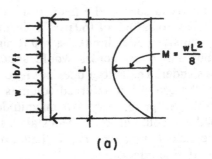

(a)

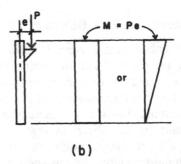

(b)

FIGURE 4.9. Bending moment in walls: (a) from wind or soil pressure; (b) from eccentric compression load.

roof level. In these cases, a bending moment is created with a magnitude equal to the product of the load and its distance of eccentricity from the centroid of the wall cross section. The following example illustrates such a situation.

Example 4. Assume that the supported load of 12,000 lb in Example 2 is placed on a bracket attached to the wall so that the center of the loading is 8 in. from the center of the wall. Investigate the condition for combined bending and axial stresses.

Solution: The average unit compressive stress was determined in Example 2 to be 34.0 psi, assuming the effective wall segment of 9 in. by 54 in. The bending moment is

$$M = Pe = (12,000)(8) = 96,000 \text{ in.-lb}$$

The section modulus for the 9-in. by 54-in.

wall segment is

$$S = \frac{bd^2}{6} = \frac{(54)(9)^2}{6} = 729 \text{ in.}^3$$

and the maximum bending stress is

$$f_b = \frac{M}{S} = \frac{96,000}{729} = 131.7 \text{ psi}$$

The limiting combined stresses are thus

$$f = f_a + f_b = 34.0 \pm 131.7$$

$$= 165.7 \text{ psi in net compression}$$

$$= 97.7 \text{ psi in net tension}$$

(See case 3 in Fig. A.7.)

The investigation indicates that the allowable compression stress of 160 psi is slightly exceeded. More critical, however, is the tension stress. Table 4.3 yields allowable tension stress in flexure of 20 psi or 40 psi, depending on the job conditions regarding inspection (as described in the code). Even if the higher value is permitted, the eccentric loading produces a situation of considerable overstress. Remedial design considerations in this case include the possibilities for modifying the support details to reduce the amount of eccentricity of the load, using a pilaster at the concentrated load (as described in Sec. 4.8), or development of the construction with steel reinforcement, as described in the next section.

The load eccentricity described in Example 4 is typically produced by the use of supporting devices attached to the face of a wall, such as that shown in Fig. 4.10a. The advantage of this type of construction is that it generally permits an undisturbed, continuous development of the wall, with only minor accommodation for anchor bolts or other anchoring devices. The disadvantage is in the degree of bending in-

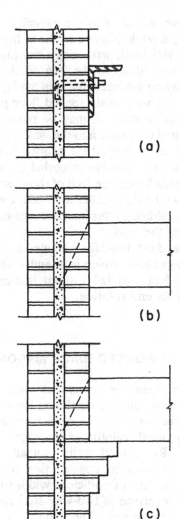

FIGURE 4.10. Beam supports in brick walls.

duced when the supported load is of considerable magnitude.

In earlier times, with thick masonry walls, it was common to use a recessed pocket in the wall to permit supported beams or ends of trusses to slightly penetrate the wall in order to get closer to the wall center and avoid the bending moments due to an eccentric load. This technique is not used as often today due to increased use of reinforced or veneered forms of construction.

Figure 4.10*b* shows a possibility for creating a pocket in one wythe of the two-wythe brick wall, producing a bearing area about 5 in. deep in the 9-in.-thick wall. This may be sufficient for light loads, such as those from closely spaced floor joists. For heavier loads, it may be possible to use a steel unit built into the wall (called a beam box) to spread the bearing load. It may also be possible to corbel (progressively cantilever) the face of the masonry, as shown in Fig. 4.10*c*, to create a wider bearing, although this increases the eccentricity of the load.

A standard detail for pocketed beams is the fire-cut end, which is intended to permit the beam to fall without tearing the wall by its end rotation.

4.7   REINFORCED BRICK MASONRY

Brick masonry is sometimes used with steel reinforcement, emulating the nature of reinforced concrete or the more frequently used reinforced masonry with CMUs. Reinforced walls typically consist of two-wythe construction with an internal cavity of sufficient width to permit the insertion of two-way steel rods in the cavity space. The cavity is then filled with concrete as the construction proceeds.

A minimum wall is usually approximately 9 in. thick, assuming a wythe thickness of 3.75 in. and a cavity width of 1.5 in. The cavity dimension is a bare minimum to permit the steel bars to pass each other, if they are as large as No. 5 bars (approximately 0.625 in. diameter).

Various code requirements must be satisfied if the construction is to qualify as reinforced in the codes' definitions of the class of construction. However, reinforcement can also be used simply for enhancement for local stress conditions,

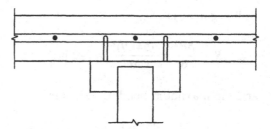

FIGURE 4.11. Plan of the reinforced wall at the beam.

even with what is otherwise "unreinforced" construction.

Consider the situation of Example 4 in the preceding section, in which an excess of bending was determined to be created by the eccentric load. If some steel reinforcement is inserted in the grouted cavity at the location of the load, as shown in the wall plan section in Fig. 4.11, it is possible to consider the resistance to flexural tension as being developed solely by the steel, which is traditionally assumed in design of reinforced concrete. The following example demonstrates such a design process.

Example. Redesign the wall for the conditions described in Example 4 in Sec. 4.6, using vertical steel reinforcement in the wall cavity to develop resistance to the flexural tension. (See Fig. 4.13.)

Solution: The vertical steel rods will tend to reinforce the wall for both the bending and the vertical compression. Since the wall without the steel was found to be almost adequate for the combined compression stresses in Example 4, the assistance provided by the steel for column action will certainly suffice in this regard. What remains is to consider the tension. Actually, the steel will be slightly precompressed by the vertical compression, but for a conservative design we will ignore this effect.

Using the bending moment determined in the previous example, a value of 20,000 psi for allowable tension stress in the steel, an effective depth of 4.5 in. (assuming the rods in the center of the wall), and the basic working-stress formula for flexural tension in reinforced concrete, we find

$$A_s = \frac{M}{f_s jd}$$

$$= \frac{96,000}{(20,000)(0.9)(4.5)} = 1.185 \text{ in.}^2$$

A conservative choice would be three No. 6 (approximately 0.75 in. diameter; see Table 2.2) rods, providing a total area of

$$A_s = (3)(0.44) = 1.32 \text{ in.}^2$$

Reinforced masonry is developed more frequently with CMUs. Design of reinforced construction is demonstrated more extensively in Chapter 5, although many of the procedures used there are generally applicable to brick construction as well.

4.8 MISCELLANEOUS BRICK CONSTRUCTION

Brick can be used for many purposes, other than for walls of buildings. Past and current uses include foundation walls, short retaining walls, freestanding fences, columns, pedestals, and fireplaces and chimneys. Construction may be all brick, or, when only some faces are exposed to view, a composite of brick and other materials: CMUs, concrete, or stone. This section considers some of these uses for brick masonry.

Columns

Built-up brick columns usually begin with the smallest size, which consists of two bricks per layer, as shown in Fig. 4.12a. The column side dimension in this case will be determined by the brick width dimension plus the mortar joint thickness. There is no real opportunity for installation of vertical steel rods, so this is strictly an unreinforced member; add the small size, and it is not likely to be used for major loads.

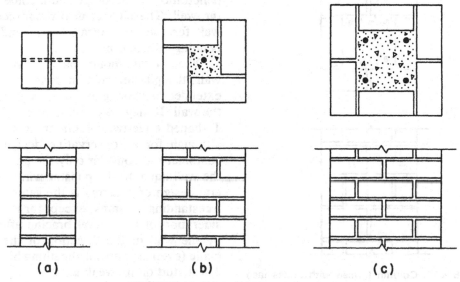

(a)   (b)   (c)

FIGURE 4.12. Forms of brick columns.

A pinwheel arrangement of the courses will produce the column shown in Fig. 4.12*b*, resulting in a small center cavity. A single steel rod might be placed in this cavity, but to create a real reinforced member it is usually necessary to use the arrangement shown in Fig. 4.12*c*. The minimum, officially reinforced brick column is therefore about 16 in. wide.

A slightly narrower reinforced column can be produced if the bricks are laid with the narrow side (face) down, as shown in Fig. 4.13. This form may be structurally acceptable, but appearance may be questionable, since bricks are usually produced with the intent of having their faces exposed, and the flat sides may have a significantly cruder finished surface.

Unreinforced columns are generally questionable, unless of very stout proportions (height-to-width ratio of, say, 10 or less). A significant usage, however, may be for pedestals, which are actually very short columns (usually defined by codes as having a height of less than three times their width).

Design of unreinforced brick columns

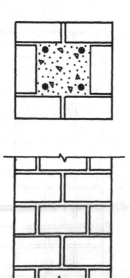

FIGURE 4.13. Column formed with bricks face down.

and pedestals is generally accomplished with the same procedures used for walls, as demonstrated in the examples in Sec. 4.6. Design of reinforced columns is done with code-specified procedures, generally adapted from those used with reinforced concrete.

A special column is the pilaster, consisting essentially of a bulged-out portion of a continuous wall. It may occur at a corner, at the edge of an opening, at the end of a wall, or at some intermediate point in a wall. The functional purposes for pilasters are usually the provision of some reinforcement for the otherwise relatively thin wall. They may serve to receive some large, concentrated load or simply to brace the wall when slenderness is a problem. Very tall walls may be braced with closely spaced pilasters, so that the unbraced distance for the wall becomes the pilaster spacing, rather than the wall height.

A frequent use of pilasters is that of receiving a load at the wall face, such as that described in Example 4 in Sec. 4.6. The portion of the pilaster extending beyond the wall face can be used to support the load, thus eliminating the need for penetration of the supported member into the wall. The pilaster also reinforces the wall for the concentrated load and any bending effects it produces.

When built monolithically with the wall, the pilaster column consists of the extended portion plus some segment of the wall. It may be possible to consider a T-shaped effective column cross section, although for a conservative design it is customary to consider only the portion of the wall equal to the pilaster width. General design of pilasters is the same as for freestanding columns, except for the consideration of the lateral bracing afforded by the wall in the direction of the wall plane (even as the wall sheathing braces a 2 × 4 stud on its weak axis).

Large brick columns, as well as large

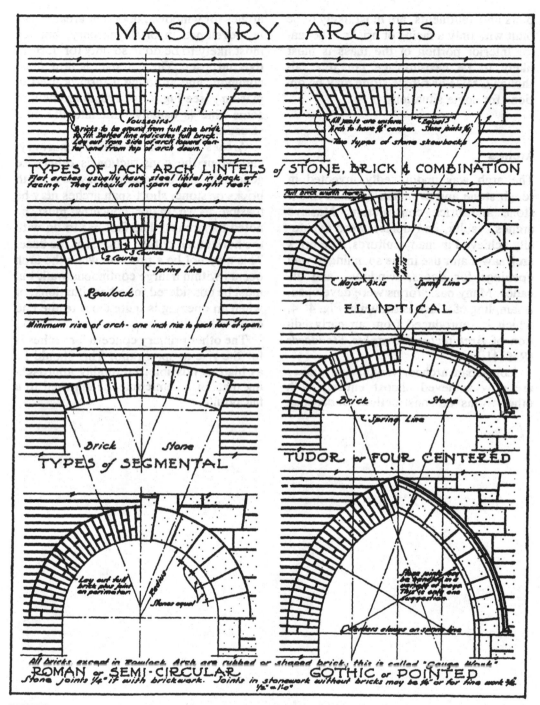

FIGURE 4.14. Forms of masonry arches in walls. Reprinted from *Architectural Graphic Standards*, 3rd edition, 1941, with permission of the publishers, John Wiley & Sons.

piers and abutments, are most likely to be built with only a facing of bricks. The major interior portion of the mass is most likely to be cast-in-place concrete, although CMUs or rubble stone may be options for some situations.

Arches

The arch was undoubtedly invented in rough stone and eventually refined in cut stone and brick construction. Massive arches, vaults, and domes were eventually achieved in many cultures. The brick arch saw major use in the spanning of wall openings for doors, windows, and arcades. Many basic forms were developed, a sampling of which is shown in Fig. 4.14, which is reproduced from an early edition of *Architectural Graphic Standards* (Ref. 11.).

Long-span arches, vaults, and domes are now achieved almost entirely with other forms of construction. Short-span arches for windows, doors, or arcades can be developed in brick masonry, but are most likely to be done so only for decorative purposes. The flat span is now most common, and although a form is used in masonry (see Fig. 4.14) a reinforcement of some kind is usually built into the construction.

Arches can still be built of masonry, and when the construction is otherwise generally of structural masonry with bricks or stone, short-span arches may be feasible. Two major concerns must be noted for arches. The first has to do with the horizontal, outward thrust at the base. This may not be a problem when the arch occurs within a large continuous wall, but must be considered for other situations, or when an opening is quite close to the end of a wall.

The other primary concern for arches is for the rise-to-span ratio. This should be as high as possible, preferably approaching that for a semicircle, as opposed to a flat profile.

5

MASONRY WITH CONCRETE UNITS

Units of precast concrete have been used for masonry construction for many years. Only in relatively recent times, however, have they been generally accepted for construction exposed to view, and been developed for major structural applications. In earlier times major use was made of hollow units of fired clay; first for utilitarian use as structural backup for veneers and plastered partitions, and later as architectural terra-cotta for decorative uses. Many of the techniques and details developed with clay tile construction were carried over and adapted with CMU construction. Today most structural masonry is achieved with precast concrete units, mostly for economic reasons but also because of the increased use of reinforced forms of construction. This chapter deals with various considerations of the use of precast concrete units for structural masonry. Many of the structural elements discussed here could also be achieved in either brick or stone construction as well,

but are much more likely to be done with CMUs.

5.1 TYPES OF CONCRETE MASONRY UNITS (CMUs)

Concrete is used for a vast array of masonry units, ranging from simple, solid bricks to complex forms used for decorative construction. The workhorse of structural construction, however, is the standard unit, occurring in one of the two forms illustrated in Fig. 2.4. This unit is standard not only in form, but in its dimensions, which are based on a three-dimensional 8-in. module. Thus the standard unit is one module high on the face, yielding a modular 8-in. course when combined with a mortar joint that is 0.5 in. or 0.375 in. thick. It is one module deep (actually only 7.5 in. or 7.625 in., which corresponds to the actual dimension of a structural lumber 2 × 8, which can be placed on the top of a wall for attachment of wood rafters). The standard unit is two modules long, combining with a vertical mortar joint for a 16-in.-horizontal-length module on the wall face. With a running bond arrangement of units (see Fig. 4.3a), the two-module long blocks overlap by one 8-in. module.

The form of the horizontal cross section of CMUs is usually developed so that the hollow cores of units in one course will line up with those in the course below when the units are laid up in the running-bond pattern. This results in vertical, tubular voids in the wall, permitting the insertion of steel reinforcing rods that can be bonded into the construction by filling the voids with lean concrete.

Vertical alignment of the voids may also permit the installation of insulative fill materials to enhance the thermal resistance of the wall. In some cases, wiring or piping may be installed in the voids to avoid the necessity to place such items on the face of the construction.

While generally conforming to some industry-wide specifications, individual manufacturers produce unique block shapes and sizes, relating to regional usage concerns, local construction prac-

tices, and local building code requirements. While the 8 × 8 × 16 in. nominal block size is a standard, units are also generally available in other sizes, including the following variations:

Thickness (wall or wythe) of 4, 6, 8, 10, and 12 in., nominal

Height (course) of 4, 6, and 8 in., nominal

Length of 8 in. (one module), 16 in. (two modules), and sometimes 12 in. (one and one-half modules), nominal

The average weight of CMU construction varies with the percentage of void created by the hollow spaces and the unit density of the concrete. Table 5.1 gives some properties for average construction with various wall thicknesses and concrete densities. Filling the voids with reinforcement and concrete will increase the average wall weight. The specific products of individual manufacturers will also vary slightly from these average values. For various items of design information, it is best to determine the particular suppliers for the region of any design project and to obtain specific data directly from the likely sources of the materials. CMUs

Table 5.1 AVERAGE PROPERTIES OF HOLLOW CONCRETE BLOCK CONSTRUCTION

Nominal Block Thickness (in.)	Net Area of Block (in.^{2} per ft of wall length)	Wall Weight in psf of Wall Surface, Where Density of Concrete in Block in lb/ft^3 Is					Moment of Inertia, I, in in.^{4} and Section Modulus, S, in in.^{3}, Where Mortar Is			
							Face-Shell Bedded[a]		Fully Bedded[b]	
		60	80	100	120	140	I	S	I	S
4	28	14	18	22	27	31	38	21	45	25
6	37	20	26	33	40	46	130	46	139	50
8	48	24	32	40	47	55	309	81	334	88
10	60	28	37	47	56	65	567	118	634	132
12	68	34	45	55	67	78	929	160	1063	183

[a] Horizontal mortar joint only at face shells.
[b] Horizontal mortar joint on full unit cross section.

Source: Adapted from table in NCMA-TEK Publication 2A, *Concrete Units*, with permission of the publisher, National Concrete Masonry Association.

are much too heavy and bulky to be transported any great distance for a construction project.

For work to be exposed to view, many variations of form, face detail, color, and block type are available. Our concerns here are principally structural, so we will not treat this issue in any depth.

Many aspects of CMU construction depend on a basic distinction as to whether the form of construction is qualified as reinforced or unreinforced, using the building codes' definitions. For some further discussion of units, details of construction, and design considerations, we will treat these two forms of construction separately. The next two sections deal with unreinforced construction, and Sec. 5.4 deals with reinforced construction.

This distinction of the structural character of the basic construction is of critical concern for design of structural masonry. However, many fundamental concerns for masonry construction in general relate to all forms of construction. Thus, concerns for material quality of units, class of mortars, provisions for crack control, incorporation of insulation, appearance of exposed faces of walls, care in the construction processes, and many other factors are always present. Quality of the general construction is a continual concern; although it becomes more of a structural concern to some degree with unreinforced construction.

5.2   UNREINFORCED CONSTRUCTION

The standard form of unit used most often for unreinforced forms of construction is that shown in Fig. 5.1a. This is called a three-cell unit, although it actually has two half-cells at each end; thus the module of repetition is four cells per block. The actual dimensions shown here—$\frac{3}{8}$ in. less than the modular dimensions—are

frequently used, anticipating slightly thinner mortar joints. This is done partly to allow for interfacing with brick construction, which sometimes occurs in multiwythe work with bricks of modular height (see Fig. 2.2c).

Staying within this standard unit module, there are typically some special units provided for achieving common elements of the construction. As shown in Fig. 5.1, these include the following:

End and Corner Units (Figs. 5.1b and c). These are used at the ends and corners of walls, where the ends of the units are exposed. The half-blocks (one 8-in. module) are used to develop the typical running-bond wall face pattern of units (see Fig. 4.3a).

Sash Units (Fig. 5.1d). These are one (as shown) or two module units with squared ends and a groove to facilitate the anchoring of window or door frames at the vertical edges of openings.

Lintel or Bond Beam Units (Fig. 5.1e). These are used over openings or at the tops of walls to create cast-in-place concrete with steel reinforcement in the form of a reinforced concrete beam internal to the masonry construction.

Pilaster Units (Fig. 5.1f). These are used to form pilaster columns within the modular system of the wall construction. The two types of units shown are alternated in vertical courses to maintain the running-bond pattern on the flat face of the wall. Units of various size are available to create different size pilasters. The units shown here would normally be filled with concrete and steel reinforcement to create a cast-in-place concrete column inside the masonry.

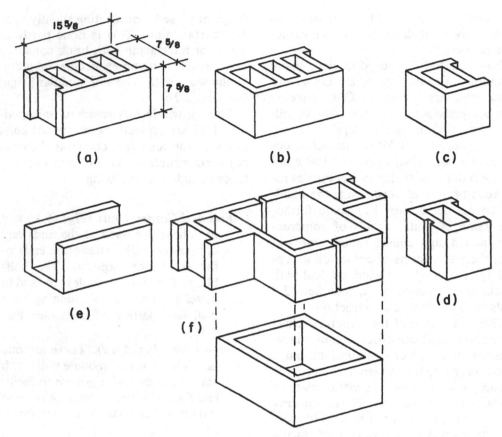

FIGURE 5.1. Forms of CMUs for unreinforced construction.

For simple structural functions, resisting compression, bending or shear, the hollow construction with concrete units is treated much the same as the solid brick construction, as illustrated in the examples in Chapter 4. The exception here is that the construction is *not* solid, and the wall cross section used for the computations must be the net section, with the voids from the full area of the wall's outer dimensions subtracted. This net section must be determined specifically from the suppliers of units, although approximate investigations can be made using the average values provided in Table 5.1. The following example demonstrates some simple cases.

Example. CMU construction is to be used for a single-wythe wall of nominal 8 in. thickness. The wall is 10 ft high and sustains on its top an axially applied, uniformly distributed load of 2000 lb/ft of wall length. Type N, ASTM C-90 units with $f_m = 1350$ psi are to be used. Investigate the wall for average maximum compressive stress. (Note: Type N is required for weather exposure; f'_m of 1350 psi is the typical assumed maximum code limit in the absence of tests on units.)

Solution: From Table 5.1, we note that the nominal 8-in.-thick wall has an average net cross section of 48 in.^{2}/ft of wall length. Thus the applied load produces a stress of

$$f_a = \frac{P}{A} = \frac{2000}{48} = 41.7 \text{ psi}$$

To this must be added the stress due to the weight of the wall, which has a maximum value at the base of the wall. Without the unit density of the concrete in the CMUs, we will assume the maximum value in Table 5.1 (140 lb/ft^3), yielding a weight of 55 lb/ft of wall height. Thus the 10-ft wall weights 10(55) = 550 lb/ft of wall length, the stress due to its own weight is

$$f_a = \frac{W}{A} = \frac{550}{48} = 11.5 \text{ psi}$$

and the total stress is 41.7 + 11.5 = 53.2 psi.

From Table 4.3, we find the maximum allowable stress for hollow-unit masonry with Type S mortar (the minimum mortar for weather-exposed structural work) to be 150 psi. The construction is therefore quite adequate.

If bending occurs in combination with axial compression, the combined stresses must be investigated for both the maximum net compression and any possible net tension. For flexural stresses we may use the section properties given in Table 5.1 (S for flexural stress). Note that Table 4.3 provides very low allowable tension stress in flexure, generally discouraging any actual net tension.

Walls must also be investigated for slenderness effects and concentrated loadings, following the procedures illustrated in the examples in Chapter 4. Unreinforced piers, columns, and pedestals can also be designed for these procedures.

Although the typical unreinforced class of construction is considered as a simple homogeneous material, permitting simple stress calculations for most loadings, investigation is somewhat more complex when some voids are grouted solid and possibly have steel reinforcement. This is often done partially, even in what is generally qualified as unreinforced construction. Procedures for these situations are described more fully in the next section.

Some procedures are generally followed in all construction with hollow concrete units, whether the work is intended for structural purposes or not. Horizontal joint reinforcement is commonly used to reduce cracking and absorb the considerable stress due to shrinkage of the mortar. Bond beams (grout-filled with steel rods) are often used for the top course. These practices simply ensure better construction. Other enhancements may relate to direct concerns for structural actions.

5.3 REINFORCEMENT FOR UNREINFORCED CONSTRUCTION

Taken in its broader context, reinforcement means anything that adds strength to the otherwise unadorned construction. As discussed in other situations in this book, this includes the use of steel elements in mortar joints or in grout-filled cavities, but it also refers to the use of pilasters and stronger masonry units at strategic locations.

Without fully transforming construction with hollow masonry units into the reinforced class, it is possible to use either stronger units or some reinforcement to give significant improvement in strength at various critical points. The general means for improving strength of hollow-unit masonry are the following:

1. Use denser concrete. This generally results in increased unit strength in the material. Some strong lightweight concretes can be achieved, but usually the lower the density, the weaker the concrete.

2. Use units with thicker walls. Typically, units made with dense concrete are intended for major structural use and sometimes have thicker walls than those made with significantly less dense material and generally intended for less demanding structural applications. The combination of denser material and thicker walls can make a considerable difference in strength.

3. Fill voids with grout (generally a slightly runny concrete). This can be done with or without adding steel bars; the concrete in the voids generally produces equivalent solid construction, which constitutes a considerable gain in wall strength by itself.

4. Use larger and/or stronger units at strategic locations. Pilasters are one example of this, and are often used at points of concentrated loads. The increased local size alone may be significant, but the units used to form pilasters are usually either thicker and stronger themselves or are formed to provide for a cast-in-place reinforced concrete column of significant size.

5. Add steel elements (bars or heavy wire) to mortar joints or grout-filled voids. This amounts to building the equivalent of reinforced concrete beams or columns into the masonry construction. This may be done to add local required strength or simply to tie the construction together. Reinforced bond beams are ordinarily used along the tops of walls for the latter purpose, even in construction that is essentially nonstructural.

Reinforcement may have a specific purpose in many cases, but is also frequently done simply to improve the general integrity of the construction. Many of the classic, ornamental details of masonry construction, handed down from ancient times, were initially developed from experience and the pragmatic concerns for producing better construction.

5.4 REINFORCED CONSTRUCTION

Hollow-unit masonry that qualifies for the building codes' definition of reinforced is typically made with standard units that provide larger voids. Thus the vertical alignment of voids provides for the forming of a reinforced concrete column of some greater dimension. Instead of the three-core unit typically used for unreinforced forms of construction (see Fig. 5.1), the standard unit used for reinforced construction is the two-core unit, as shown in Fig. 5.2. Actually, as discussed in Sec. 5.2, the units shown in Fig. 5.1 have a four-core module, due to the half core at each end. Thus the units used for reinforced construction actually have voids that are close to twice in size to those in unreinforced construction.

With fewer voids, there are fewer cross walls in the units, and the general strength of the masonry is slightly less in reinforced forms of construction. Compensating significantly for this is the existence of the cast-in-place reinforced concrete

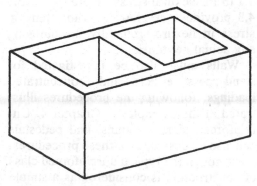

FIGURE 5.2. Basic form of the two-cell CMU for reinforced construction.

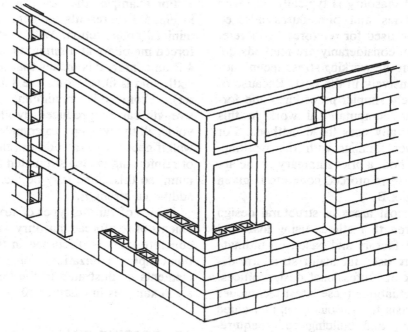

FIGURE 5.3. Form of the internal reinforced concrete structure in reinforced construction with CMUs.

column and beam rigid frame that is developed inside the masonry, as shown in Fig. 5.3. This frame does the double job of tying the masonry together into an integral mass and developing its own independent strength as a reinforced concrete structure.

Building code requirements ensure the presence of the vertical and horizontal reinforced members at least every 4 ft. Additional elements are specifically required at the tops and ends of walls and around all sides of openings. The result, shown in Fig. 5.3, is a rather extensive framework of reinforced concrete. The overall strength of the construction derives significantly from this encased structure and often somewhat minimally from the encasing masonry.

There are, in fact, priority systems that use units placed without mortar, which derive their structural integrity almost entirely from the encased structure. However, most reinforced construction is done

with structurally rated units, laid with mortar of high grade and generally producing a masonry construction with some significant integrity. For some minor structural tasks, such as may occur with low-rise buildings with short-span roof systems, the masonry itself may be sufficient for general structural purposes; in which case the encased reinforced concrete structure is a bonus.

To a large extent, the fully reinforced forms of hollow-unit masonry now in use were developed in response to needs for structures with enhanced resistance to the effects of earthquakes and windstorms. The poor response of unreinforced construction in these situations is well documented and is, unfortunately, repeatedly demonstrated in almost every major earthquake and windstorm. Building codes in regions at high risk to these disasters now generally prevent use of unreinforced construction.

Structural investigation and design of

reinforced masonry is typically achieved with theories and procedures adapted from those used for reinforced concrete. Still in use considerably are methods deriving from the working-stress techniques used extensively in the past. Because of its relatively simpler procedures and formulas, the computational work in this book generally uses these methods. For readers not familiar with the working-stress method, a brief summary of the applications in reinforced concrete is given in Appendix B.

As in most areas of structural design today, strength methods using load and resistance factors and based on ultimate failures are used for major structural design work in professional design offices. In most situations these latest techniques are the basis for various computer-aided procedures, and building code requirements tend toward an acknowledgment of this situation. Still, for design of minor work and for simple structures in general, considerable use is made of tabulations in various references and some of the very simple, approximate procedures of the working-stress method.

Minimum Construction

As in other types of construction, building codes and industry standards result in a certain minimum form of construction. The structural capacity of this minimum construction represents a threshold that may be sufficiently high to provide for many situations of structural demand. Such is frequently the case with reinforced construction, so quite frequently, for buildings of modest size, the construction developed in simple response to code minimum requirements is more than adequate. In any event, the minimum construction represents a take-off point for consideration of any modifications that may be required for special structural tasks.

For example, the construction shown in Fig. 5.3 represents the general form of minimal construction. With vertical reinforced members at a minimum spacing of 4 ft on center, this means that only every sixth void is filled (voids are 8 in. on center). From this minimum, increased strength can be produced by filling more voids, all the way up to a completely filled wall if necessary. In addition, the amount of reinforcing is also stipulated as a minimum, so this alone may be increased for additional strength.

Design of various types of structural elements of reinforced masonry with hollow concrete units is discussed in the rest of this chapter. Utilization of many of these elements is illustrated in the building design examples in Chapter 10.

5.5   BEARING WALLS

Walls made with CMUs are often used to support gravity loads from roofs, floors, or walls in upper stories. Walls serving such functions are called bearing walls. The principal structural task is resistance to vertical compression force, which may involve considerations for one or more of the following:

> *Average Compressive Stress.* This is simply the total compression load (including the wall weight) divided by the area of the wall cross section. Care should be taken to indicate whether this is the average stress for the gross section, defined by the wall's outer dimensions, or the net section, which is the actual solid portion of the hollow units.
>
> *Bearing Stress.* This is the actual contact pressure under a load that is concentrated, such as that under the end of a supported beam.
>
> *Effective Column Stress.* This is the stress considered when the wall sup-

ports widely spaced, concentrated loads, and only a limited part of the wall is considered effective.

Bending Stress. This may be produced when the vertical load is not applied in an axial manner to the wall, such as when a beam is not placed on top of a wall, but is supported by a bracket or ledger on the face of the wall, resulting in bending plus compression loadings.

Investigations for stress conditions in solid brick walls for all of these conditions are demonstrated in the examples in Chapter 4. For construction with unreinforced hollow units, the procedures are essentially the same, except for consideration of the voided cross section of the wall. See Example 1 in Sec. 5.2.

Walls may also sustain concentrated loads, as illustrated in Example 2 in Sec. 4.6 for a brick wall. The following example demonstrates a similar procedure for a CMU wall.

Example 1. A CMU wall sustains concentrated loads of 12,000 lb at 8-ft centers. The wall is of 8 in. nominal (7.625 in. actual) thickness, with blocks of concrete with density of 100 lb/ft^3, producing a wall with approximate weight of 40 psf of wall surface. Blocks are approximately 50% solid, f'_m for the construction is 1350 psi, and the wall has an unbraced height of 10 ft. The load is placed axially on the wall through a steel bearing plate that is 6 in. by 16 in. Investigate the wall for bearing and compression stresses.

Solution: For the calculated bearing stress we find

$$f_{br} = \frac{P}{A} = \frac{12,000}{6 \times 16} = 125 \text{ psi}$$

which is compared to the allowable bearing of

$$F_{br} = 0.26f'_m = 0.26(1350) = 351 \text{ psi}$$

The bearing stress was calculated without reduction for the hollow-wall cross section, since it is assumed that the bearing will be on a fully grouted bond beam at the top of the wall. However, in this example, the stress would not be critical even with the reduced effective area.

For the maximum compressive stress we consider the situation at the bottom of the wall, for which the wall weight must be added to the applied load. We find the wall weight to be

$$W = 40(10)$$

$$= 400 \text{ lb/ft of wall length in plan}$$

Using an effective wall pier of six times the wall thickness or four times the wall thickness plus the bearing width, we use 6(7.625) = 45.75 in. or 4(7.625) + 16 = 46.5 in., say 46 in. The total weight of this pier is therefore

$$W = \frac{46}{12}(400) = 1533 \text{ lb}$$

Thus the total load is 12,000 + 1533 = 13,533 lb, and the average compressive stress in the pier is

$$f_a = \frac{P}{A_e} = \frac{13,533}{(7.625)(46)(0.50)} = 77.2 \text{ psi}$$

This must be compared with the allowable stress, which is typically taken from Table 4.3 for the unreinforced construction. For this case, the limiting value is 150 psi, and the construction is adequate.

Unreinforced walls may sustain reasonably heavy loads, as long as no bending is present. Bending may be added from the effects of eccentricity of the vertical load or from lateral loads due to wind, earthquakes, or soil pressures. This results in a combined stress condition that may be critical for either the net total compression or the net tension, where such is present. (See discussion in Sec. A.8 in Appendix A.)

Details of the construction often make it difficult to place supported loads axially on walls (load center over the wall center), especially with multistory or parapet walls. The following example illustrates the effect of such a condition on an unreinforced wall.

Example 2. Suppose that the load in Example 1 is placed on the wall face, as shown in Fig. 5.4, so that an eccentricity of 6.8125 in. occurs between the load center and the wall center. Investigate the wall for the combined stress condition.

Solution: With the eccentricity as shown in Fig. 5.4, the load develops a bending moment of

$$M = Pe = (12,000)(6.8125)$$

$$= 81,750 \text{ in.-lb}$$

For the bending stress, we find from Table 5.1 that the wall has a section modulus (S)

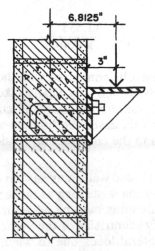

FIGURE 5.4.

of 88 in.³/ft of wall. For the 46-in. pier the total value is

$$S = \frac{46}{12} (88) = 337 \text{ in.}^3$$

and the maximum bending stress is

$$f_b = \frac{M}{S} = \frac{81,750}{337}$$

$$= 243 \text{ psi}$$

Since this alone is in excess of the allowable stress of 150 psi from Table 4.3, there is no need to add it to the axial compression. Although the allowable compression is exceeded, the more critical concern is for the net tension, which will exceed 170 psi near the top of the wall (where the wall weight will not reduce it).

Clearly, an unreinforced wall cannot develop resistance to this magnitude of bending. Efforts must be made to reduce the bending effects or to make other provisions for it. Reduction can be made by moving the load closer to the wall center. (See Fig. 4.10 for a brick wall.) Other provisions may consist of adding steel reinforcement or a pilaster column.

Reinforced walls can sustain both greater axial compression and higher ranges of bending. Grouting of some void spaces to install the reinforcement adds to the wall cross section. For concrete block walls, the grouted spaces can usually be assumed to be at least as strong as the concrete units. Thus a fully grouted wall may be considerably stronger for direct compression.

Various methods can be used for design of reinforced walls. We use the working-stress method here because it is simple, but it will often produce quite conservative results. The following examples demonstrate the use of the methods discussed in Appendixes A and B.

Example 3. Suppose the wall in Example 2 is developed with reinforced construction. Using steel with $F_y = 40$ ksi, investigate the wall.

Solution: The general condition that must be satisfied for the reinforced wall is

$$\frac{f_a}{F_a} + \frac{f_b}{F_b} \leqslant 1$$

In this formula, f_a and f_b are the calculated axial and bending stresses, and F_a and F_b are the corresponding allowable stresses. Each value can be determined, and the combination can be evaluated. However, a first condition worth investigating is whether the wall can sustain the moment at all. This is determined by calculating the maximum resisting moment as Kbd^2. (See discussion in Sec. B.4.) For the wall with $f'_m = 1350$ psi and steel with $F_y = 40$ ksi, Table B.2 yields $K = 66.6$ (in in.-lb units). Considering the "beam" width to be the full 46-in. width of the pier, and $d = 7.625/2 = 3.8125$ in. with reinforcing in the center of the wall, the maximum bending moment without any axial compression is

$$M_R = Kbd^2 = (66.6)(46)(3.8125)^2$$

$$= 44{,}530 \text{ in.-lb}$$

which indicates that the moment in Example 2 cannot be sustained, even without consideration for the combined stresses.

Adding grouted void spaces and some vertical steel reinforcement does indeed increase wall strength, but not without limits. In many situations the major improvement to the general construction when reinforced masonry is used is the overall toughness gained by the structure.

Example 4. Consider the wall in the preceding examples to be subjected to a uniform load of 2500 lb/ft and a lateral wind

force of 20 lb/ft^2 on the wall surface. Investigate the wall (a) as an unreinforced wall and (b) as a reinforced wall of minimum construction.

Solution: For the wind pressure, the wall spans vertically as a simple beam on the 10-ft span, producing of maximum bending moment of

$$M = \frac{wL^2}{8} = \frac{(20)(10)^2}{8}$$

$$= 250 \text{ ft-lb, or } 250(12)$$

$$= 3000 \text{ in.-lb}$$

(a) Unreinforced Wall

Investigating for combined compression and bending stresses, we will consider the situation of the wall at approximately mid-height, where the bending moment is maximum. From Table 5.1, assuming a density of 100 lb/ft^3 for the concrete and fully bedded mortar joints, the properties of the wall are

Average weight = 40 psf of wall surface

Average net cross section of wall = 48 in.^{2}/ft of length

Average section modulus $(S) = 88$ in.^{3}/ft of wall length

At mid-height we add to the applied load a wall weight of

$$W = \text{(unit weight)(half the wall height)}$$

$$= (40)\left(\frac{10}{2}\right) = 200 \text{ lb}$$

The total compression load at mid-height is thus $2500 + 200 = 2700$ lb, and the average net compression stress is

$$f_a = \frac{P}{A_e} = \frac{2700}{48} = 56.25 \text{ psi}$$

The maximum bending stress is

$$f_b = \frac{M}{S} = \frac{3000}{88} = 34.09 \text{ psi}$$

and the combined stresses are

$$\text{Maximum } f = 56.25 + 34.09$$
$$= 90.34 \text{ psi}$$
$$\text{Minimum } f = 56.25 - 34.09$$
$$= 22.16 \text{ psi (compression)}$$

This indicates that there is no net tension stress, and the maximum value for compression is well below the limit of $150(1.33) = 200$ psi as given in Table 4.3 and increased for wind loading.

The allowable stress based on slenderness (h/t) of the wall should also be investigated, and the possible effect of interaction should be considered $(f_a/F_a + f_b/F_b \leqslant 1)$. This will not be critical in this case. The procedure for this is demonstrated in the following investigation of the reinforced wall.

(b) Reinforced Wall

To qualify as reinforced, the wall must satisfy various criteria, including the following:

Horizontal and vertical reinforcement in grouted voids, maximum of 48 in. spacing

Minimum bar size, No. 4

Minimum percentage of A_s, sum of both directions, $0.002A_g$ of wall

Minimum percentage of A_s, either direction, $0.0007A_g$ of wall

The minimum total reinforcement (sum of both ways) is thus

$$A_s = (0.002)(7.625)(12) = 0.183 \text{ in.}^2/\text{ft}$$

Favoring the horizontal direction due to

increased shrinkage effects, a possible choice is therefore (see Table C.1)

Horizontal: No. 6 at 48 in., $A_s = 0.110$ in.^{2}/ft

Vertical: No. 5 at 48 in., $A_s = 0.077$ in.^{2}/ft

Based on the minimum vertical reinforcement, the total moment resistance can be expressed as (See Appendix Sec. B.4.)

$$M_R = A_s f_s j d$$
$$= (0.007)(20,000 \times 1.33)(0.9)(3.8125)$$
$$= 7028 \text{ in.-lb}$$

For the masonry, flexural compression is limited to

$$F_b = 0.333 f'_m = (0.333)(1350)(1.333)$$
$$= 600 \text{ psi}$$

which must be reduced by 50% if special inspection and testing are not provided.

Using the lower value, we can express the total moment resistance as

$$M_R = \tfrac{1}{2}(F_b k j b d^2)$$
$$= \tfrac{1}{2}[(300)(0.33)(0.9)(12)(3.8125)^2]$$
$$= 7770 \text{ in.-lb}$$

Axial compression will be approximately the same as for the unreinforced wall. The grouted construction weighs slightly more, but the net wall cross section is slightly increased. Referring to the calculation for the unreinforced wall, we will assume an average value of $f_a = 60$ psi.

For the allowable axial compression stress, we use the formula based on wall slenderness:

$$F_a = 0.20 f'_m \left[1 - \left(\frac{h'}{42t} \right)^3 \right]$$

$$= 0.20(1350)\left[1 - \left(\frac{120}{42 \times 7.625}\right)^3\right]$$
$$(1.333)$$

$$= 340 \text{ psi (or 170 psi without inspection)}$$

The investigation for the interaction condition is thus as follows:

$$\frac{f_a}{F_a} + \frac{f_b}{F_b} = \frac{60}{170} + \frac{3000}{7046}$$
$$= 0.353 + 0.426 = 0.779$$

which indicates a safe condition.

Note that we have substituted a ratio of actual bending moment to limiting bending moment for the f_b/F_b term. This is possible because f_b is produced by the actual moment and F_b by the limiting moment; thus the ratio of the moments is the same as the ratio of the stresses.

These calculations indicate that the wall is adequate with or without reinforcement. In fact, there seems to be no real gain in strength by addition of the reinforcement. This is due partly to the limitations of the working-stress method; use of strength methods will usually indicate more capacity for the reinforced construction. However, the real gains achieved by the two-way reinforcement are truly more in the form of the enhanced toughness of the general construction.

Bearing walls frequently also serve other functions, so their final design must relate to their full utilization. Some of the other major wall functions are considered in this chapter.

5.6 BASEMENT WALLS

Basement walls usually perform an earth-retaining function. They also often function as vertical load-carrying bearing walls, spanning grade beams, or as the base for building shear walls. A complete design must include considerations for all of the wall requirements.

For their earth-retaining functions, basement walls ordinarily span vertically between levels of horizontal support. For a single-story basement the support at the bottom of the wall is provided by the edge of the concrete basement floor slab, and the support at the top of the wall is provided by the first floor structure of the building. If an active soil pressure of the equivalent fluid type is assumed, the load and the structural actions for the wall will be as shown in Fig. 5.5a when the exterior ground level is close to the top of the wall. When the ground level is below the top of the wall, the pressure is as shown in Fig. 5.5b, and the lateral bending effect is considerably reduced.

If heavy vehicles can be driven close to the side of a building, or the finished grade level is actually above the top of the basement wall, a surcharge effect will occur, as shown in Fig. 5.5c. If the basement is multilevel, the wall may function as a multiple span element for the pressure condition shown in Fig. 5.5d.

For buildings of light construction with relatively short basement walls, it is common to use construction of masonry or concrete with no reinforcement. Precise investigation of many such walls would indicate some overstress by current code requirements, but long experience with few failures is generally accepted as a valid reason for continuing the practice. Nevertheless, when the wall spans more than 10 ft or so or is subjected to surcharge effects, the use of vertical reinforcement is strongly advised.

For light residential construction the form of masonry most often used is that with CMUs. For best results the following minimum construction is recommended:

Concrete units with minimum density of 100 pcf

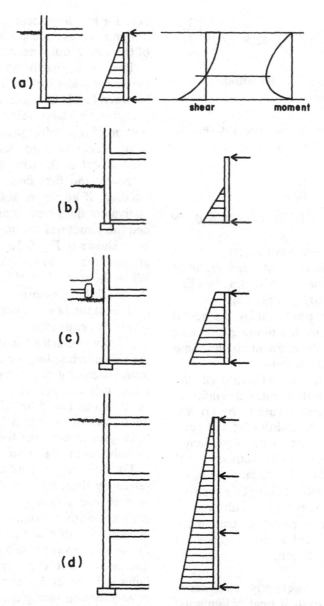

FIGURE 5.5. Various situations for lateral pressure on basement walls.

Type M mortar, joints fully bedded

Void spaces fully grouted

Top course developed as a reinforced bond beam

There are many general requirements for basement walls. A critical concern is that for water penetration, the potential for which depends on the specific groundwater situation. When the groundwater level is well below the bottom of the wall, the need is usually limited to that for dampproofing. This is achieved by the ensuring of good construction with tight mortar joints and all the usual methods to avoid cracking and, typically, by the addition of a water-repelling coating on the outside surface of the wall below the ground level.

If a high groundwater condition exists, or extensive irrigation of landscaping materials near the building is anticipated, it is necessary to consider the use of waterproofing, which is generally equivalent to that required for a flat roof. This usually requires the use of more extensive surfacing materials as well as special treatment of wall joints and basement floors to obtain fully water-sealed construction.

Structural design of basement walls is essentially the same as that for other walls subjected to combinations of vertical compression plus bending. The following example illustrates the process for a short wall with CMU construction.

Example. A basement wall is to be built as shown in Fig. 5.6, using CMUs with density of 120 lb/ft^3, f_m = 1350 psi, and 50% voids. Assuming no surface surcharge and an equivalent fluid pressure of 30 psf for the soil, investigate the wall.

Solution: The loading conditions for the wall are shown in Fig. 5.6b. These include a vertical loading of 1500 lb/ft on top of the wall and the lateral force due to soil pressure as shown. The triangular form of distributed load produces a maximum moment of

$$M = 0.128PL = (0.128)(1500)(10)$$

$$= 1920 \text{ ft-lb or } 23,040 \text{ in.-lb}$$

With fully bedded mortar joints the section modulus of the wall is (see Table 5.1)

$$S = 132 \text{ in.}^3/\text{ft of wall}$$

and the maximum bending stress is

$$f_b = \frac{M}{S} = \frac{23,040}{132} = 175 \text{ psi}$$

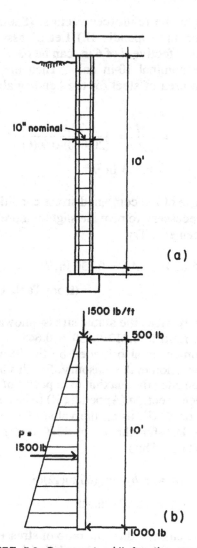

FIGURE 5.6. Basement wall for the example: (a) form and dimensions; (b) loading condition.

This value of stress exceeds the maximum of 170 psi from Table 4.3, so it is not necessary to consider the additional stress due to axial compression. The wall is clearly unacceptable without some vertical reinforcement.

If vertical reinforcement is used, it should be placed as close as possible to the inside of the wall, rather than in the center of the wall, because this results in a more favorable value for the effective depth of the reinforced section. (See discussion in Appendix B.) Let us assume that an effective d of 6 in. can be obtained in the nominal 10-in units. Then the required area of steel for the bending alone is

$$A_s = \frac{M}{f_s j d} = \frac{23,040}{(20,000)(0.9)(6)}$$

$$= 0.213 \text{ in.}^2/\text{ft}$$

Because of the combined stress condition, it is necessary to provide slightly more reinforcement. Try

No. 6 at 16, $A_s = 0.330 \text{ in.}^2/\text{ft}$

(from Table C.1)

which reduces the actual stress–allowable stress ratio to 0.213/0.330 = 0.645.

Moment is also limited by the flexural compression in the masonry. For this limit we consider the maximum capacity of the balanced section (Appendix B) to be a moment of Kbd^2, using the value of K from Table B.1 for the f'_m of 1350 and f_s of 20,000 psi. Thus

$$M = Kbd^2 = (66.6)(12)(6)^2$$

$$= 28,771 \text{ in.-lb}$$

On the basis of this, the ratio of stresses is the ratio of the actual and limiting moments:

$$\frac{f_b}{F_b} = \frac{23,040}{28,771} = 0.801$$

which becomes the critical value.

For the vertical compression the maximum stress will occur at the bottom of the wall. However, the greater concern is for the combined compression and bending condition, which is more critical near the mid-height of the wall. For a reasonable combination we will consider the total vertical compression load with the applied load of 1500 lb/ft plus the wall weight at 7 ft below the top of the wall. Thus

Wall weight (fully grouted)

$$= \left(\frac{9.5}{12}\right)(120)(7) = 665 \text{ lb}$$

Total load = 1500 + 665 = 2165 lb

The maximum average compressive stress is

$$f_b = \frac{P}{A} = \frac{2165}{(9.5)(12)} = 19 \text{ psi}$$

This is a very low stress for the reinforced construction, and it should be obvious that the combined loading is not critical. The choice of the No. 6 bars at 16 in. is quite conservative, and a strength design procedure would probably indicate that No. 5 at 16 is adequate.

It is also possible, of course, to increase the wall thickness to a nominal 12 in., which might make an unreinforced wall an alternative choice.

An old rule of thumb is that the basement wall thickness in inches should not be less than the wall height in feet. This would indicate that the 10-in. wall is just barely adequate in this case, although its true thickness is just under the limit.

5.7   RETAINING WALLS

Retaining walls serve to retain the lateral pressure of soil. The basement wall is thus one form of retaining wall. However, the term is most often used to refer to a cantilever retaining wall, which is a freestanding structure without lateral support at its top. For such a wall the major design consideration is for the height of the wall, as this generally establishes the degree of difference in soil levels on the two sides of the wall and the magnitude of the lateral effect of soil pressure.

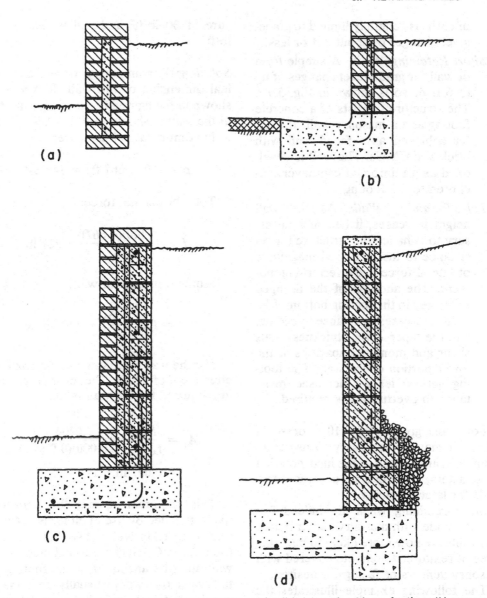

FIGURE 5.7. Forms of cantilever retaining structures: (a) tall brick curb with no footing; (b) masonry curb with gutter footing of cast concrete; (c) short retaining wall with brick face bonded to CMU backup; (d) CMU wall of graduated thickness.

Based largely on the difference in elevation of the soil surface on the two sides of a retaining structure, some different categories are as follows (see Fig. 5.7):

Curbs. These are very short structures, often used at the edges of walks or drives or around planting areas. Figure 5.7a shows a form for a brick curb placed without a foundation. A possible variation is shown in Fig. 5.7b, where a sitecast concrete element is used to achieve a foundation and to form a gutter. Use of curbs is generally limited to grade-level changes of about 2 ft or less.

Short Retaining Walls. A simple form of wall for grade-level changes of up to 6 ft or so is shown in Fig. 5.7c. The structure consists of a concrete footing and a masonry wall, the latter achieved as a faced wall with brick and CMUs. The wall is developed as a reinforced cantilever, anchored to the footing.

Tall Retaining Walls. As the wall height increases, it becomes necessary to enhance the structure for resistance to the increased magnitude of lateral force and overturning moment. The addition of the dropped shear key in the footing bottom (Fig. 5.7d) increases resistance to sliding, and the tapered wall thickness adds shear and moment capacities at the lower portion of the wall. The footing gets wider as increased resistance to overturning is required.

For walls greater than 10 ft or so in height, it may be necessary to use additional measures, such as the incorporation of pilasters, buttresses, or transverse walls for lateral bracing.

Short retaining walls are indeed frequently made with masonry construction. Tall walls are now more likely to be structures of reinforced concrete covered with masonry veneers or facings, if desired.

The following example illustrates the process for design of a short retaining wall of the form shown in Fig. 5.7c.

Example. A retaining wall is proposed with the form shown in Fig. 5.8. The masonry wall is to be achieved with concrete units with density of 120 lb/ft^3 and f'_m of 1350 psi. Investigate the structure for adequacy of the masonry, required reinforcement for the wall, and stability against sliding and overturn. Use lateral soil pressure of 30 lb/ft^2 and soil weight of 100 lb/ft^3.

Solution: For simplicity, we use the nominal dimension of the wall thickness, as shown in the figure. The loading condition in the wall is shown in Fig. 5.8b.

Maximum lateral pressure:

$$p = (30)(4.667 \text{ ft}) = 140 \text{ psf}$$

Total horizontal force:

$$H_1 = \frac{(140)(4.667)}{2} = 327 \text{ lb}$$

Moment at base of wall:

$$M = (327)\left(\frac{56}{3}\right) = 6104 \text{ in-lb}$$

For the wall we assume an approximate effective d of 5.5 in. The tension reinforcing required for the wall is thus

$$A_s = \frac{M}{f_s j d} = \frac{6104}{(20,000)(0.9)(5.5)}$$
$$= 0.061 \text{ in.}^2/\text{ft}$$

This requirement may be conservatively satisfied by use of either No. 4 at 32 (0.075 in.^{2}/ft) or No. 5 at 48 (0.077 in.^{2}/ft). (See Table C.1.) The vertical steel in the wall must be anchored to the footing. In tall walls this would usually be done by providing dowels, equal in size and at the same spacing as the wall bars. For the short wall of the form shown in Fig. 5.7c, a single bar may be used, bent into an L-shaped form, and cast into the footing. Depending on the orientation of the footing to the wall, this single bar may be used to reinforce the cantilevered toe of the footing as well.

For an investigation of stress in the masonry, we may consider the balanced moment capacity of the wall, using the low

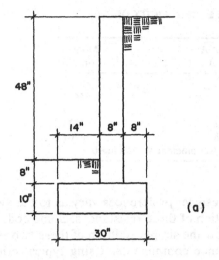

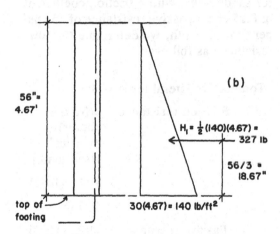

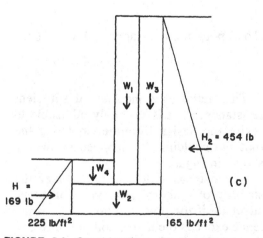

FIGURE 5.8. Considerations for the retaining-wall example.

value of masonry strength for a situation not involving special inspection (value of f_m reduced to half). Thus from Table B.2, $K = 33.3$, and the maximum resisting moment based on the compressive strength of the masonry in flexure (Appendix Sec. B.4) is

$$M_R = Kbd^2 = (33.3)(12)(5.5)^2$$
$$= 12,088 \text{ in.-lb}$$

Because this is considerably greater than the previously determined moment of 6104 in.-lb due to the soil pressure, the wall is adequate.

The loading condition used to investigate the soil stresses and the stress conditions in the footing is shown in Figure 5.8c. In addition to the limit of the maximum allowable soil-bearing pressure, it is usually required that the resultant vertical force be kept within the kern limit of the footing. The location of the resultant force is therefore usually determined by a moment summation about the centroid of the footing plan area, and the location is found as an eccentricity from this centroid.

Table 5.2 contains the data and calculations for determining the location of the resultant force that acts at the bottom of the footing. The position of this resultant is found by dividing the net moment by the sum of the vertical forces as follows:

$$e = \frac{5560}{1082} = 5.14 \text{ in.}$$

For the rectangular plan area of the footing (1 ft by 2.67 ft) the kern limit is one sixth of the footing width, or 5.33 in. (See Appendix Sec. A.8.) The load resultant is thus within the kern, and the combined stresses are

$$P = \frac{N}{A} \pm \frac{M}{S}$$

Table 5.2   DETERMINATION OF THE ECCENTRICITY OF THE RESULTANT FORCE

Force (lb)	Moment Arm (in.)	Moment (in.-lb)
$H_2 = 454$	22	+9988
$w_1 = 350$	4	−1400
$w_2 = 333$	0	0
$w_3 = 311$	12	−3732
$w_4 = 88$	8	+704
Sum = 1082 lb	Net moment = 5560 in.-lb	

where

N = total vertical force = 1082 lb

A = plan area of footing strip = 2.67 ft^2

M = net moment about the footing centroid = 5560 in.-lb

S = section modulus of the rectangular footing strip area, determined as

$$S = \frac{bh^2}{6} = \frac{(1)(2.67)^2}{6} = 1.188 \text{ ft}^3$$

The limiting maximum and minimum soil pressures are thus

$$\text{Maximum } p = \frac{N}{A} + \frac{M}{S}$$

$$= \frac{1082}{2.67} + \frac{5560/12}{1.188}$$

$$= 405 + 390 = 795 \text{ psf}$$

Minimum p = 405 − 390 = 15 psf

The maximum pressure is quite low, so the design is probably adequate. If a design soil pressure of at least 1000 psf is not possible, it is doubtful that any structure should be built on the soil.

Resistance to horizontal sliding is offered by a combination of the friction on the bottom of the footing and the lateral passive pressure against the buried portion of the wall on the low side. Codes and designer preferences vary as to the evaluation of these resistances. Some codes allow the simple addition of these two resistance components. Using typical criteria for sandy soils, with a friction coefficient of 0.25 and a passive resistance of 150 psf per foot of depth, we determine the total resistance as follows:

Total active lateral force = 454 lb

Friction resistance = [(friction factor) (vertical dead load)]

$\qquad$ = (0.25)(1082)

$\qquad$ = 270 lb

Passive resistance = $\frac{1}{2}$[(225)(1.5)]

$\qquad$ = 169 lb

Total potential resistance = 270 + 169

$\qquad$ = 439 lb

This indicates some lack of sufficient resistance, so it is probably advisable to revise the design by either lowering the footing or adding the dropped shear key shown in Figure 5.7d.

In most cases designers consider the stability of a short cantilever wall to be adequate if the potential horizontal resistance exceeds the active soil pressure and the resultant of the vertical forces is within the kern of the footing. However,

Table 5.3 ANALYSIS FOR OVERTURNING EFFECT

Force (lb)	Moment Arm (in.)	Moment (in.-lb)
Overturn		
$H_2 = 454$	22	9988
Restoring Moment		
$w_1 = 350$	20	7000
$w_2 = 333$	16	5328
$w_3 = 311$	28	8708
$w_4 = 88$	8	704
		Total = 21,700

the stability of the wall is also potentially questionable with regard to the usual overturn effect. If this investigation is considered to be necessary, the procedure is as follows.

The loading condition is the same as that used for the soil stress analysis and shown in Figure 5.8c. As with the vertical soil stress analysis, the force due to passive soil resistance is not used in the moment calculation since it is only a potential force. For the overturn investigation the moments are taken with respect to the toe of the footing. The calculation of the overturning and the dead load restoring moments is shown in Table 5.3. The safety factor against overturn is determined as follows:

$$SF = \frac{\text{restoring moment}}{\text{overturning moment}}$$

$$= \frac{21,740}{9988} = 2.17$$

The overturning effect is usually not considered to be critical as long as the safety factor is at least 1.5.

Table 5.4 gives design data for short reinforced masonry retaining walls varying in height from 2 to 6 ft. Table data have been developed using the procedures illustrated in the design example. Details and

criteria for the walls are shown in Fig. 5.9. Note that the illustration shows two necessary conditions. The first concerns the profile of the ground surface behind the wall. If this is excessively steep, there will be a significant surcharge effect on the wall. Table designs are based on the assumption of a relatively flat profile, up to a maximum slope of 1 : 5.

A second concern for the wall involves the possible buildup of water in the soil behind the wall. This should be avoided by using a reasonably fast-draining fill (gravel) and by placing drains through the wall as shown.

Wall thicknesses in Table 5.4 are based on typical nominal block sizes. For the determination of the wall weight, it is assumed that the blocks are made with medium-density concrete and that all voids are filled with normal-density concrete.

5.8 SHEAR WALLS

Structural masonry walls are quite often used as shear walls, forming part of the lateral-load-resisting system for a building. Various systems can be used for lateral bracing, but the most common one for low-rise construction is the box system, typically formed with a combination of horizontal and vertical diaphragms.

Table 5.4 SHORT MASONRY RETAINING WALLS[a]

Wall Height, H (ft-m)	Wall		Footing			Reinforcing			Actual Maximum Soil Pressure (psf-kPa)
	Nominal t (in.-mm)	Assumed Weight (psf-kPa) of Wall Surface	w (in.-mm)	h (in.-mm)	A (in.-mm)	1	2	3	
2–0.6	6–140	55–2.6	18–450	6–150	4–100	No. 3 at 48–1200	—	2 No. 3	55–26
2.67–0.8	6–140	55–2.6	22–550	6–150	6–150	No. 3 at 32–800	—	2 No. 3	600–29
3.33–1.0	8–190	75–3.6	27–675	8–200	8–200	No. 4 at 48–1200	No. 4 at 48–1200	2 No. 4	700–34
4–1.2	8–190	75–3.6	32–800	10–250	10–250	No. 4 at 32–800	No. 4 at 32–800	3 No. 4	850–41
4.67–1.4	8–190	75–3.6	40–1000	12–300	12–300	No. 4 at 24–600	No. 3 at 24–600	4 No. 4	850–41
5.33–1.6	0–240	95–4.6	48–1200	14–350	15–375	No. 4 at 24–600	No. 4 at 24–600	5 No. 5	825–40
6–1.8	0–240	95–4.6	56–1400	16–400	18–450	No. 5 at 24–600	No. 4 at 24–600	5 No. 5	850–41

[a] See Figure 5.1).

Criteria: Active pressure = 30 lb/ft²

Passive pressure = 150 lb/ft²

Coefficient of friction = 0.25

Concrete: f'_c = 2000 lb/in²

Steel: f_s = 20,000 lb/in²

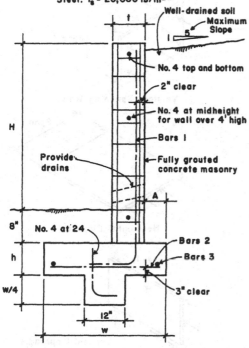

FIGURE 5.9. Reference figure for Table 5.4.

Framed deck systems of the roof and floors are assumed to act as rigid planes (called diaphragms), receiving the edge loading from the windward wall and distributing the loading to the vertical bracing elements.

Vertical frames or shear walls, acting as vertical cantilevers, receive the loads from the horizontal diaphragms and transfer them to the building foundations.

The foundations anchor the vertical bracing elements and transfer the loads to the supporting ground.

The propagation of the loads through the structure is shown on the left in Fig. 5.10, and the functions of the major elements of the lateral resistive system are shown on the right. The exterior wall on the windward side functions as a simple spanning element loaded by a uniformly distributed pressure normal to its surface and delivering a reaction force to its supports. Even though the wall construction may be continuous through several stories, it is usually considered as a simple span at each story level, thus delivering half of its load to each support. According to Fig. 5.10, this means that the upper wall delivers half of its load to the roof edge and half to the edge of the second floor. The lower wall delivers half of its load to the second floor and half to the first floor.

The roof and second-floor diaphragms function as spanning elements loaded by the edge forces from the exterior wall and spanning between the end shear walls, thus producing a bending that develops tension on the leeward edge and compression on the windward edge. It also produces shear in the plane of the diaphragm that becomes a maximum at the end shear walls. In most cases the shear is assumed to be taken by the diaphragm, but the tension and compression forces due to bending are transferred to framing at the dia-

Horizontal diaphragms ordinarily consist of the roof and framed floor constructions, which constitute rigid, horizontal planar elements. The horizontal diaphragm collects forces and distributes them to the vertical elements of the bracing system. Figure 5.10 shows a simple rectangular building under the effect of wind load normal to one of its flat sides. The complete lateral resistive system that responds to this loading consists of the following:

Wall surface elements on the windward side are assumed to take the direct wind pressure and are typically designed to span vertically between the roof and floor structures.

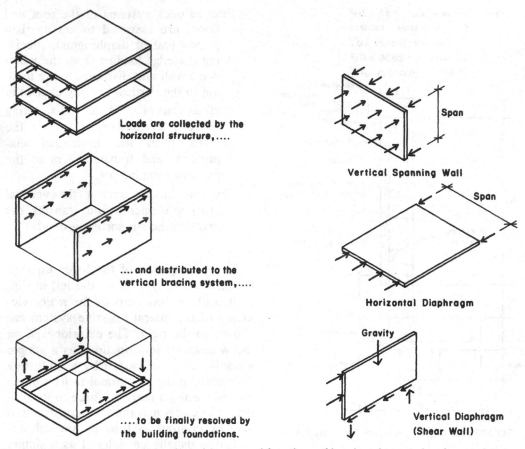

Loads are collected by the horizontal structure,....

....and distributed to the vertical bracing system,....

....to be finally resolved by the building foundations.

Vertical Spanning Wall

Span

Horizontal Diaphragm

Span

Vertical Diaphragm (Shear Wall)

Gravity

FIGURE 5.10. Propagation of wind forces and functions of bracing elements in a box system.

phragm edges. The means of achieving this transfer depends on the materials and details of the construction.

The end shear walls act as vertical cantilevers that also develop shear and bending. The total shear in the upper story is equal to the edge load from the roof. The total shear in the lower story is the combination of the edge loads from the roof and second floor. The total shear force in the wall is delivered at its base in the form of a sliding friction between the wall and its support. The bending caused by the lateral load produces an overturning effect at the base of the wall as well as the tension and compression forces at the edges of the wall. The overturning effect is resisted by the stabilizing effect of the dead load on the wall. If this stabilizing moment is not sufficient, a tension tie must be made between the wall and its support.

If the first floor is attached directly to the foundations, it may not actually function as a spanning diaphragm but rather will push its edge load directly to the leeward foundation wall. In any event, it may be seen in this example that only three quarters of the total wind load on the building is delivered through the upper diaphragms to the end shear walls.

This simple example illustrates the basic nature of the propagation of wind forces through the building structure, but there are many other possible variations with more complex building forms or with other types of lateral resistive structural systems.

Seismic Forces

Seismic loads are actually generated by the dead weight of the building construction. In visualizing the application of seismic forces, we look at each part of the building and consider its weight as a horizontal force. The weight of the horizontal structure, although actually distributed throughout its plane, may usually be dealt with in a manner similar to the edge loading caused by wind. In the direction normal to their planes, vertical walls will be loaded and will function structurally in a manner similar to that for direct wind pressure. The load propagation for the box-shaped building in Fig. 5.10 will be quite similar for both wind and seismic forces.

The Box or Panelized System

The box or panelized system is usually of the type shown in the preceding example, consisting of some combination of horizontal and vertical planar elements. Actually, most buildings use horizontal diaphragms simply because the existence of roof and floor construction provides them as a matter of course. The other types of systems usually consist of variations of the vertical bracing elements. An occasional exception is a roof structure that must be braced by trussing or other means when there are a large number of roof openings or a roof deck with little or no diaphragm strength.

Elements of the building construction developed for the gravity load design, or for the general architectural design, may become natural elements of the lateral resistive system. Walls of the proper size and in appropriate locations may be theoretically functional as shear walls.

Whether they can actually serve as such will depend on their construction details, on the materials used, on their height-to-width ratio, and on the manner in which they are attached to the other elements of the system for load transfer. It is also possible, of course, that the building construction developed only for gravity load resistance and architectural planning considerations may *not* have the necessary attributes for lateral load resistance, thus requiring some replanning or the addition of structural elements.

The most common shear wall constructions are those of poured concrete, masonry, and wood frames of studs with surfacing elements. Wood frames may be made rigid in the wall plane by the use of diagonal bracing or by the use of surfacing materials that have sufficient strength and stiffness. Choice of the type of construction may be limited by the magnitude of shear caused by the lateral loads, but will also be influenced by fire code requirements and the satisfaction of the various other wall functions.

General Behavior of Shear Walls

As shown in Fig. 5.11, the general functions of shear walls are the following:

1. *Direct Shear Resistance*. This usually consists of the transfer of a lateral force in the plane of the wall from some upper level of the wall to a lower level or to the bottom of the wall. This results in the typical situation of shear stress and the accompanying diagonal tension and compression stresses.

2. *Cantilever Moment Resistance*. Shear walls generally work like vertical cantilevers, developing compression on one edge and tension on the opposite edge, and transferring an overturning moment to the base of the wall.

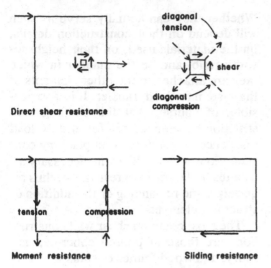

FIGURE 5.11. Functions of a shear wall.

3. *Horizontal Sliding Resistance.* The direct transfer of the lateral load at the base of the wall produces the tendency for the wall to slip horizontally off its supports.

The shear stress function is usually considered independently of other structural functions of the wall. The maximum shear stress that derives from lateral loads is compared to some rated capacity of the wall construction, with the usual increase of one third in allowable stresses because the lateral load is most often a result of wind or earthquake forces. For concrete and masonry walls the actual stress in the material is calculated and compared with the allowable stress for the material.

The moment effect on the wall is usually considered to be resisted by the two vertical edges of the wall acting as flanges or chords. In the concrete or masonry wall this results in a consideration of the ends of the wall as columns, sometimes actually produced as such by thickening of the wall at the ends. In wood-framed walls the end-framing members are considered to fulfill this function. These edge members must be investigated for possible critical combinations of loading because of gravity and the lateral effects.

The overturn effect of lateral loads must be resisted for the building as a whole, as well as for individual elements of the vertical lateral bracing system. For wind, the overturning caused by the horizontal forces must be combined with the uplift caused by upward suction pressure on the roof. For buildings with a height-to-width ratio of 0.5 or less and a maximum height of 60 ft, the combination of the effects of overturning and uplift may be reduced by one third. Weight of earth over footings may be used to calculate the dead-load-resisting moment. For both the entire building and its individual lateral bracing elements, the overturning moment must not exceed two thirds of the dead-load-resisting moment [see *UBC* Sec. 2311(e)].

For seismic effects, *UBC* Sec. 2312(h)1 specifies that only 85% of the dead load be used to resist uplift effects when using the working-stress method for materials. This means that any anchorage elements that are required can be designed for their working-stress resistance.

For an individual shear wall, the overturn investigation is summarized in Fig. 5.12. The figure shows a single direction load, but the wall must be designed for load from either direction. The dead load indicated may be due only to the wall weight, but typically includes the dead weight of some supported structure as well.

Resistance to horizontal sliding at the base of a shear wall is usually at least partly resisted by friction caused by the dead loads. For masonry and concrete walls with dead loads that are usually quite high, the frictional resistance may be more than sufficient. If it is not, shear keys must be provided. For wood-framed walls the friction is usually ignored, and the sill bolts are designed for the entire load.

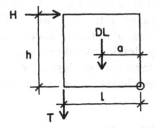

To determine T:
  for wind —   DL(a) + T(l) = 1.5 [H(h)]
  for seismic — 0.85 [DL(a)]+ T(l) = H(h)

FIGURE 5.12. Determination of stability and tiedown requirements for a shear wall; working-stress method (*Uniform Building Code*).

Design and Usage Considerations

An important judgment that must often be made in designing for lateral loads is that of the manner of distribution of lateral force between a number of shear walls that share the load from a single horizontal diaphragm. In some cases the existence of symmetry or of a flexible horizontal diaphragm may simplify this consideration. In many cases, however, the relative stiffnesses of the walls must be determined for this calculation.

If considered in terms of static force and elastic stress–strain conditions, the relative stiffness of a wall is inversely proportionate to its deflection under a unit load. Figure 5.13 shows the manner of deflection of a shear wall for two assumed conditions. In (*a*) the wall is considered to be fixed at its top and bottom, flexing in a double curve with an inflection point at midheight. This is the case usually assumed for a continuous wall of concrete or masonry in which a series of individual wall portions (called *piers*) are connected by a continuous upper wall or other structure of considerable stiffness. In (*b*) the wall is considered to be fixed at its bottom only, functioning as a vertical cantilever.

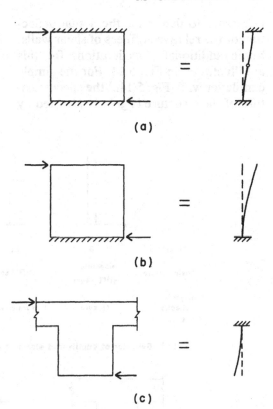

FIGURE 5.13. Forms of deflection of shear walls: (*a*) fully fixed wall, top and bottom; (*b*) cantilever wall, fixed base; (*c*) cantilever wall, fixed top.

This is the case for independent, free-standing walls or for walls in which the continuous upper structure is relatively flexible. A third possibility is shown in (*c*) in which relatively short piers are assumed to be fixed at their tops only, which produces the same deflection condition as in (*b*).

In some instances the deflection of the wall may result largely from shear distortion, rather than from flexural distortion, perhaps because of the wall materials and construction or the proportion of wall height to plan length. Furthermore, stiffness in resistance to dynamic loads is not quite the same as stiffness in resistance to static loads.

For various purposes it is sometimes

necessary to determine the actual deflection or the relative stiffness of shear walls. Some additional considerations for this are illustrated in Fig. 5.14. For the simple cantilever wall (Fig. 5.14a) the specific nature of the structure is greatly affected by its height-to-length ratio, h/l (relating to span-to-depth ratio in beams). The ordinary flexing of a beam occurs mostly in relatively slender members, with h/l ratios of 10 or more, which typically occur with rafters and joists. At most, a shear wall

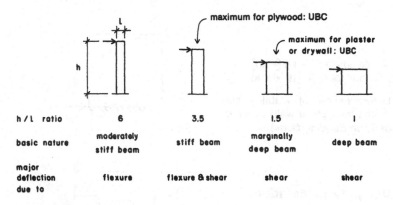

h / L ratio	6	3.5	1.5	1
basic nature	moderately stiff beam	stiff beam	marginally deep beam	deep beam
major deflection due to	flexure	flexure & shear	shear	shear

(a) Behavior of cantilevered elements related to height-to-length ratios

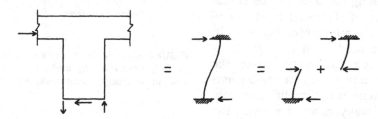

(b) Deflection assumption for a fully fixed masonry pier

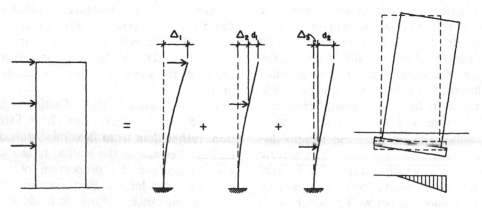

(c) Deflection of a multistory shear wall

(d) Shear wall tilt caused by uneven soil pressure

FIGURE 5.14. Aspects of lateral deflection of shear walls.

may act as a stiff beam, but it more frequently acts as a deep beam, in which shear distortion is the principal mechanism producing deflections.

Deflections may be simple in nature (e.g., for the single cantilever wall) or compound, as shown for the doubly fixed pier in Fig. 5.14*b* or the multistory wall in Fig. 5.14*c*. If deflection computation is necessary, these compound deformations may be broken down to simpler components, as shown in the figure.

A major concern for vertical bracing elements is the possibility of yielding or other loss of full fixity at the supports. For the cantilever wall resting on a simple, shallow bearing footing, the uneven distribution of soil pressure may cause some significant rotation at the base. Although possibly not resulting in actual overturn failure, this can still represent a significant loss of bracing stiffness, resulting in transfer of load to other parts of the building construction.

Use of Stiffness Factors

A common situation that occurs is that in which a number of individual shear walls or piers share some total lateral force, requiring that the distribution of force to the individual bracing elements be determined. Figure 5.15 shows three situations in which this can occur for a single wall.

In Fig. 5.15*a* a wall is formed by a series of separated, but linked, masonry walls, with lighter construction forming the wall portions between the masonry. If the individual masonry piers are all of the same size and similarly constructed, the total lateral force in the wall will be simply divided equally between the piers. If they have different dimensions, however, their relative stiffnesses must be used to apportion the load to the piers.

For the wall in Fig. 5.15*b* a similar situation

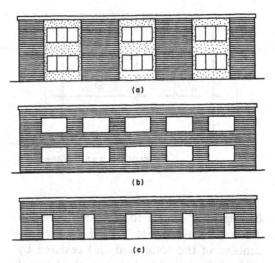

FIGURE 5.15. Form variations for masonry shear walls: (*a*) individual, isolated, linked piers (vertical cantilevers with fixed bases); (*b*) continuous wall with fully fixed, individual piers; (*c*) continuous wall with individual cantilever piers, fixed at their tops.

ation occurs if the distribution of shear on a horizontal plane through the window openings is considered. In this case the individual piers of masonry between the window openings act as fully fixed elements, as shown in Fig. 5.13*a*.

For the piers between the door openings in Fig. 5.15*c*, the condition may be one of full fixity (5.15*b*) or simple cantilever (Figs. 5.15*a* or 5.13*c*), depending on the nature of the anchorage and support at the base of the piers.

Assuming that the piers are all similarly constructed, the basic factor that distinguishes them from each other in any of the walls in Fig. 5.15 is their aspect ratio (vertical dimension to horizontal dimension). Strictly on the basis of this ratio, plus their qualification of single fixity (simple cantilever) or double fixity, their relative stiffnesses can be established and used as a basis for distribution of lateral loads. Factors for this purpose are given in the tables in Appendix C, and their use is demonstrated in the following example.

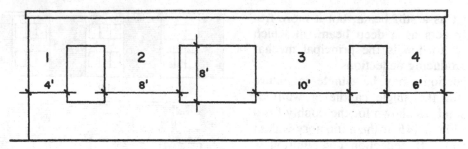

FIGURE 5.16. Multiple-pier wall for the example problem.

Example. A lateral force is delivered to the wall shown in Fig. 5.16. Find the percentage of the total load (H) resisted by each of the individual piers at the level of the window openings.

Solution: In this case the piers are considered to be fixed at their tops and bottoms. Stiffness factors, R_c, for the individual piers are thus obtained from Table C.3, on the basis of the h/d ratios for the piers. The load distribution to each individual pier is then determined by multiplying the total load by a distribution factor, DF:

$$DF = \frac{\text{factor for the individual pier}}{\text{sum of the factors for all the piers}}$$

The computations for the distribution are summarized in Table 5.5.

Construction

Just about any form of structural masonry offers some potential for development of a shear wall. Wind and earthquake forces acted on ancient buildings just as they now do on modern ones. Ancient walls still standing testify to their adequacy in this regard.

However, we now generally build much lighter and thinner masonry structures and have enchancements at our disposal to create better structures. We also have many ways to attach the separate elements of building systems to each other. And, of major significance, we generally document and disseminate data regarding our collective experiences for all to share. Although we inherited many valuable lessons from the masons and builders of the past, the art of design for lateral force effects is a fairly recent development.

Table 5.5 LOAD DISTRIBUTION TO THE MASONRY PIERS

Pier	h (ft)	d (ft)	h/d	R_c	DF[a]	Share of Lateral Load (%)
1	8	4	2.0	0.1786	0.087	8.7
2	8	8	1.0	0.6250	0.304	30.4
3	8	10	0.8	0.8585	0.417	41.7
4	8	6	1.33	0.3942	0.192	19.2
				Sum = 2.0563		

[a] $DF = \dfrac{R_c \text{ for pier}}{\text{sum of } R_c}$.

Where wind forces are moderate and earthquakes virtually unknown to have occurred, it is possible to develop shear walls with unreinforced construction. Where windstorms are prevalent or major earthquakes are a high risk, or simply where major loads must be resisted, it is now preferred to use reinforced construction. Codes provide data and procedures for both situations, and typically require only reinforced construction where severe wind or seismic effects are present.

Usage of masonry structures for lateral load resistance is illustrated in the building design cases in Chapter 10. These are simple cases using common construction elements. Structural design and development of construction details for complex structures or for very severe loading conditions must be done with the highest state of the art in terms of engineering and construction, and are beyond the scope of this book.

Some general considerations for shear wall design are the following:

1. Anchorage of walls to supports for resistance to sliding or overturn is generally achieved by doweling the reinforcement in reinforced construction. Special anchors or keys may be necessary with unreinforced construction, although the sheer weight of the wall is often sufficient for its stability.
2. Walls must be adequately supported, and bearing foundations should be very conservatively designed to minimize any soil deformations.
3. Connections of supported structures that deliver lateral loads to walls should be very carefully designed for the combined gravity and lateral loads. Positive resistance to lateral loads is critical, especially for seismic loads.

4. Special attention should be given to stress conditions at the wall discontinuities at openings, wall corners, and intersecting walls. These should be reinforced, even in otherwise unreinforced construction.

5.9 PEDESTALS

There are many instances in which a short compression element, called a pier or pedestal, is used as a transition between a footing and some supported element. Some of the purposes for pedestals are the following:

To Permit Thinner Footings. By widening the bearing area on the top of the footing, the pedestal will achieve a reduction in the shear and bending stresses in the footing, permitting the use of a thinner footing. This may be a critical issue where allowable soil bearing pressures are quite low.

To Keep Wood or Metal Elements above the Ground. When the bottom of the footing must be some distance below the ground surface, a pedestal may be used to keep vulnerable elements above ground.

To Provide Support for Elements at Some Distance above the Footing. In addition to the previous situation, there are others in which elements to be supported may be some distance above the footing. This may occur in tall crawl spaces, basements, or where the footing elevation must be dropped a considerable distance to obtain adequate bearing.

Pedestals of masonry or concrete are essentially short columns. When carrying major loads, they should be designed as reinforced columns with appropriate vertical reinforcement, ties, and dowels. When loads are light and the pedestal height is less than three times the thickness, however, they may be designed as unreinforced elements. If built of hollow masonry units (concrete blocks, etc.) they should have the voids completely filled with concrete. Table 5.6 gives data for short, unreinforced masonry pedestals. The table entries represent only a sampling of the potential sizes for these pedestals and are intended to give a general idea of the load range, not to be construed as standard or recommended sizes.

There are minimum as well as maximum heights for pedestals. If the pedestal is very short and the area of the bearing contact with the object supported by the pedestal is small, there will be considerable bending and shear in the pedestal, similar to that in a footing. This can generally be avoided if the pedestal is at least as tall as it is wide. For a pedestal of concrete it is theoretically possible to reinforce the pedestal in the same manner as a footing, although this is usually not feasible.

Masonry pedestals are subject to considerable variation. Primary concerns are for the type of masonry unit, the class of mortar, and the general type of masonry construction. For the pedestals in Table 5.6 we have assumed a widely used type of construction, utilizing hollow concrete masonry units (concrete blocks) with all voids filled with concrete. For this construction we have used the design criteria in the *Uniform Building Code* (Ref. 1) as given in Table 4.3, assuming Type S mortar. The allowable stress in compression on the gross cross-sectional area of the pedestal is 150 psi [1.03 MPa]. According to a footnote to the table, an increase of 50% is permitted for the stress at the bearing contact with the object supported by the pedestal. Based on these require-

ments, it is possible to establish a maximum allowable load for a given size pedestal, and to derive the minimum column size on the basis of the ratio of the two allowable stresses. These two items are given in Table 5.6.

For supported objects with contact area smaller than the minimum size listed, the pedestal load will be limited by the contact area rather than by the potential pedestal capacity. Thus the table includes pedestal capacities for a range of column sizes, based on the stress limit for the contact area.

Although the pedestals in Table 5.6 are designed using criteria for unreinforced masonry, we recommend the use of vertical reinforcing in the four corners of the pedestal when the pedestal height exceeds twice the pedestal thickness. This reinforcing is quite minimal in cost and need not be doweled into the footing, but it adds some degree of toughness to the pedestal against the ravages of shrinkage, temperature changes, and possible damage during construction of the supported structure. As stated previously, if the pedestal height exceeds three times its thickness, the pedestal should be designed as a reinforced masonry column with appropriate vertical reinforcement, horizontal ties, and footing dowels.

Pedestals may also be constructed with bricks, one possible advantage being the potential for increased bearing stress. However, because pedestals are mostly not exposed to view, the construction with CMUs, as shown in Table 5.6, is likely to be more economical in most situations.

Forms of construction with CMUs, other than those shown in Table 5.6 are also possible. Special units ordinarily used to form masonry columns, as discussed in the next section, can be used with or without reinforcement.

Table 5.6 UNREINFORCED MASONRY PEDESTALS

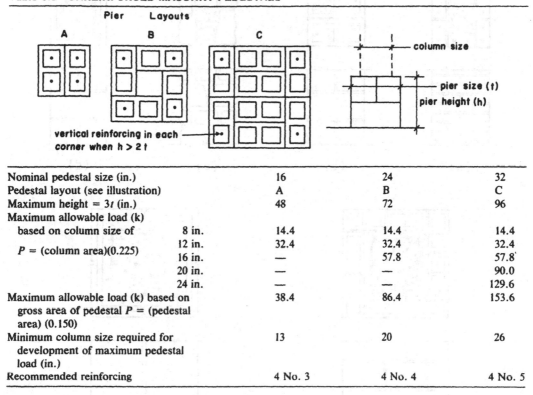

	A	B	C
Nominal pedestal size (in.)	16	24	32
Pedestal layout (see illustration)	A	B	C
Maximum height = $3t$ (in.)	48	72	96
Maximum allowable load (k)			
based on column size of 8 in.	14.4	14.4	14.4
12 in.	32.4	32.4	32.4
P = (column area)(0.225) 16 in.	—	57.8	57.8
20 in.	—	—	90.0
24 in.	—	—	129.6
Maximum allowable load (k) based on gross area of pedestal P = (pedestal area) (0.150)	38.4	86.4	153.6
Minimum column size required for development of maximum pedestal load (in.)	13	20	26
Recommended reinforcing	4 No. 3	4 No. 4	4 No. 5

5.10 COLUMNS

Masonry columns may take many forms, the most common being a simple square or oblong rectangular cross section. The general definition of a column is a member with a cross section having one dimension not less than one third the other and a height of at least three or more times its lateral dimension. Shorter columns are called pedestals (see Sec. 5.9), and elements with longer, thinner plan dimensions are considered as wall piers. See discussion in Appendix A.

The three most common forms of construction for structural columns are the following (see Fig. 5.17):

Unreinforced Masonry. These may be brick (Fig. 5.17*a*), CMU construc-

tion (Fig. 5.17*b*) or stone (Fig. 5.17*c*), and are generally limited to forms bordering on the pedestal category—that is, very stout. Slender columns, or any column required to develop significant bending or shear, should be reinforced.

Reinforced Masonry. These may be any recognized form of reinforced construction, but are mostly either fully grouted and reinforced brick construction or CMU construction with all voids filled (Figures 5.17*a*, *b* and *d*). Large columns formed with shells of CMUs are more likely in the next category.

Masonry-Faced Concrete. These are essentially reinforced concrete columns with masonry shells. The

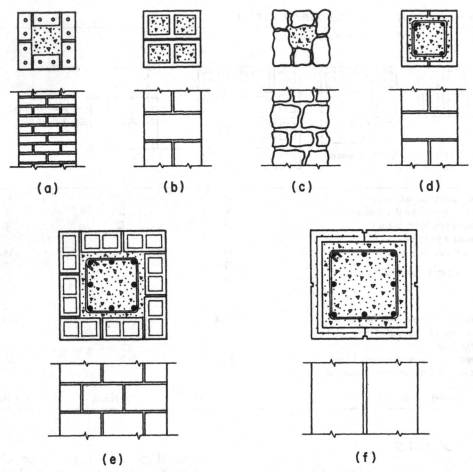

FIGURE 5.17. Forms of masonry columns: (a) brick or low-void CMUs, unreinforced; (b) CMUs with or without reinforcement; (c) stone, fully grouted; (d) CMU shell with cast concrete column; (e) large concrete column with CMU shell; (f) sitecast concrete column formed with precast concrete units (larger scale version of d).

shells may be veneers (applied after the concrete is cast) or laid up to form the cast concrete. In the latter case, the structure may be considered as a composite one, but is often more conservatively designed, ignoring the capacity of the masonry facing.

Short columns of unreinforced construction are generally designed for simple compression resistance, using the appropriate limit for compressive stress from

Table 4.3. Bearing stresses are generally allowed to be higher, but the general design of the column will not be based on bearing, unless the contact bearing area is close to the column gross cross-sectional area.

If simple compression is the major concern, the design of an unreinforced column is similar to that for a vertical load-bearing wall. (See Sec. 5.5.) If the compression load is applied with some eccentricity, it is necessary to consider the combined compression and bending stresses, as discussed in Appendix Sec. A.8.

Example 1. A short column is to be formed of CMU units as shown in Fig. 5.17*b*, with 8 × 8 × 16 in. nominal blocks having f'_m = 1500 psi. What is the maximum axial compression load for this column, if it is unreinforced and the units are laid with Type S mortar?

Solution: Assuming bearing is not critical, the maximum stress for the cross section from Table 4.3 is 150 psi for hollow-unit masonry. With the voids unfilled and assuming a 50% solid block, the total load capacity is

$$P = F_a(\text{net area}) = (150)(15.5)^2(0.50)$$

$$= 18,019 \text{ lb}$$

Reinforced masonry columns are designed as described in Appendix Sec. B.4. As with present design practice for reinforced concrete columns, it is generally accepted that all columns should be designed for some bending, with a minimum bend assumed to be that created by a load eccentricity equal to one tenth of the column side dimension. The following example demonstrates the process for a simple column.

Example 2. Assume the column in Example 1 to have its voids fully grouted and to be reinforced with four No. 5 bars with F_y = 40 ksi. Investigate the column for a compression load of 20 kips placed so that it is 2 in. off of the column center. Unbraced height is 16 ft.

Solution: If bending is ignored, the axial load capacity is limited to

$$P_a = (0.20f'_m A_e + 0.65 A_s F_{sc}) \times$$

$$\left[1 - \left(\frac{h}{42t}\right)^3\right]$$

where

$$A_e = \text{effective net area} = (15.5)^2$$

$$= 155 \text{ in.}^2 \text{ (fully grouted)}$$

$$A_s = 4(0.31) = 1.24 \text{ in.}^2$$

$$F_{sc} = \text{allowable steel stress} = 0.4F_y$$

$$= 16 \text{ ksi}$$

Thus

$$P_a = [(0.20)(1500)(155)$$
$$+ (0.65)(1.24)(16,000)] \times$$

$$\left[1 - \left(\frac{16(12)}{42(15.5)}\right)^3\right]$$

$$= 53,474 \text{ lb}$$

For the interaction relationship, therefore, f_a/F_a = 40/53.5 = 0.748, which indicates some margin for bending. For an approximate estimate of the bending capacity, we may consider the section as one with tension reinforcement only. Thus, with two No. 5 bars in the centers of the voids, the approximate effective depth is 11.6 in., and the moment limited by the steel is

$$M_R = A_s F_s jd = (0.62)(20,000)(0.9)(11.6)$$

$$= 129,456 \text{ in.-lb}$$

Comparing this with the actual moment of 40(2) = 80 kip-in., or 80,000 in.-lb, we find the ratio of f_b/F_b to be

$$\frac{80,000}{129,456} = 0.618$$

and the interaction response is thus

$$\frac{f_a}{F_a} + \frac{f_b}{F_b} = 0.748 + 0.618 = 1.366$$

Since this exceeds 1, the column is not adequate.

This is a conservative, approximate analysis, and a more accurate one using strength methods and including the effect of the compression in the other bars would probably show less overstress.

Masonry columns are also frequently developed as pilasters—that is, columns built monolithically with a wall. These may take various cross-sectional shapes. Pilasters typically serve multiple functions—reinforcing the wall for concentrated loads, excessive bending, or spanning, and bracing the wall to reduce its slenderness.

Except for the bracing afforded by the wall, pilasters are usually designed structurally as freestanding columns. This may in some cases reduce the required wall functions to that of infill between the columns.

Pilasters may be used at the ends of walls or at the edges of large openings for reinforcement. The chords for shear walls may be developed as pilasters where significant overturn is present.

6

STONE MASONRY

Prior to this century stone was used as a major structural building material. Now it is mostly either too expensive, too scarce, or too difficult to work with in comparison to alternative structural materials. However, real stone remains a popular material for building exteriors, so extensive use is made in the form of surfacing materials. This chapter deals primarily with current structural uses for stone; which is no longer the major use of the material.

6.1 RUBBLE AND FIELDSTONE CONSTRUCTION

The earliest uses of stone were undoubtedly in the form of rock piles made with stones as found in nature. Fitting the stones together to form a stable pile surely required a long patient search for the right size and shape of stone as the pile grew. Even today, stone structures are best achieved with some attention to funda-

mental principles slowly developed by skillful pilers of rocks. Some of these basic tricks, as shown in Fig. 6.1, are the following:

1. Most stones should be of a flat or angular form to minimize the tendency for upper stones to roll off of lower ones.

2. Vertical joints in successive layers (courses) should be offset rather than aligned. This helps develop a certain horizontal continuity of the structure.

3. As with a pile of any generally loose material (soil, sand, pebbles), a tapered profile with a wider base is preferred for stability. If a single-direction lateral force must be resisted, the pile may be leaned in opposition to the force. (See Fig. 6.1e.)

4. If the stack is tall, the horizontal layers should be dished slightly (Fig.

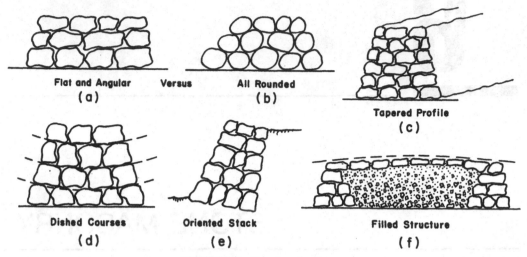

FIGURE 6.1. Good rock piles.

6.1*d*) so that the whole stack leans slightly to the inside.

5. Very wide stacks may be filled (Fig. 6.1*f*), with larger stones reserved for sides and topping. Most of the filler should be coarse, granular material (broken rock, gravel, coarse sand), but top and sides may be filled with some clay materials to seal the stack (ancient form of crude mortar).

As the ancient builders learned, the best stone structures are those that maintain a stable equilibrium without assistance. Mortar should be used primarily to fill voids after the rocks are settled in place. Once hardened, the mortar may further stabilize the pile, but it should not be used initially to prop up the stones.

Stones may be used as found, or they may be shaped. Depending on the source, natural forms may be rounded, angular, or flat. All forms may be used, but the angular and flat shapes will usually produce more stable structures. Rounded shapes may be used to fill spaces between some stones, but should not be used extensively.

Shaping may be minor and crude (just breaking or chipping), or it may be done with great precision and accuracy. Construction with natural or minor shaped stones is called rubble. Work done with stones shaped to reasonably accurate rectangular forms is called ashlar.

If stones are laid in precise horizontal layers, the work is described as coursed. If there is no particular attempt to achieve layers within the mass of the pile, the work is described as random. The general combinations of rubble, ashlar, coursed, and random are shown in Fig. 6.2. Variations and combinations are possible, and complex structures may use many forms and types of stone and great variety in arrangements.

Unreinforced Construction

Stone structures for buildings must generally be built with the same care and requirements as required for construction with bricks or CMUs. Code specifications and data can be used for work that follows good construction practices. The resulting construction can be as structurally sound as other forms of unreinforced masonry.

Stone structures will generally be somewhat thicker and heavier than those of bricks or CMUs. Joining of the stone

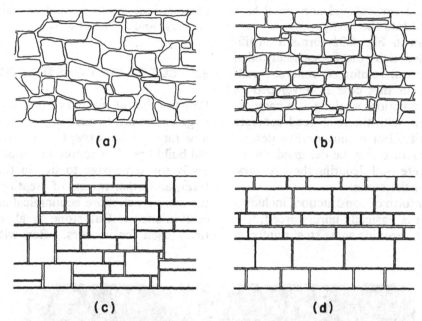

FIGURE 6.2. Patterns of stone masonry: (*a*) random rubble; (*b*) coursed rubble; (*c*) random ashlar; (*d*) coursed ashlar. Reproduced from *Fundamentals of Building Construction* (Ref. 16) with permission of the publishers, John Wiley & Sons.

work with other elements of the building construction must be achieved with details that allow for the somewhat rough dimensional accuracy of the stone construction. Installation of anchor bolts or other attachment devices may require some greater effort in random rubble construction.

For long walls some horizontal reinforcement should be used. If horizontal courses are achieved, ordinary joint reinforcement may be used. Otherwise, or as an alternative, bond beams may be developed as with CMU construction.

A bond beam should be used at the top of a wall. If a stone top is desired, the bond beam may be in the second course. If the wall provides bearing for a supported structure, however, the top should be developed as a cast concrete member for this purpose; serving to provide better bearing and to distribute the loads to the stone work.

Reinforced Construction

Stone walls that carry major loads are best developed as reinforced structures. These may take various forms, but the two shown in Fig. 6.3 are frequently used. The wall shown in Fig. 6.3*a* is developed in the same general way as a fully grouted, reinforced brick wall. A significant cavity is formed in the center of the wall and regu-

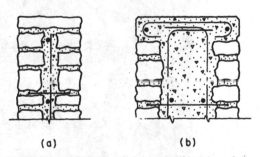

FIGURE 6.3. Reinforced stone walls: (*a*) as fully grouted (solid) masonry; (*b*) as a faced concrete wall.

larly spaced vertical and horizontal bars are grouted into the cavity.

The wall in Fig. 6.3*b* forms a significantly wider cavity that is essentially filled with a sitecast reinforced concrete wall. The stone in this case is developed as faced construction, working as a composite structure with the reinforced concrete in the cavity. For a conservative design, such a structure may be designed simply as a concrete wall, ignoring the structural capacity of the stone.

For any form of construction, including that which is basically unreinforced, it is advisable to provide some steel reinforce-

ment around openings and at wall ends, corners, and intersections.

6.2 CUT STONE CONSTRUCTION

Construction of masonry structures with large cut stones (quarried granite, etc.) is now rarely done, except in restoration of old buildings. The source material is generally too expensive to use in this bulk form, and other means of creating structures are much more economical and less craft-intensive. Cut stone is almost entirely used for veneers, often with sup-

FIGURE 6.4. Low retaining structures of stone, constructed without mortar.

porting structures of steel or reinforced concrete.

Where stone is plentiful and other technologies in limited use, it may be possible to use cut stone to generate structures, but they will generally be most feasible in the form shown in Fig. 6.3*b*. If the stone indeed forms a mold for the cast concrete, anchorage of the stone to the concrete will most likely be achieved with metal ties set into slots in the stone or into joints between the stone units. Because of the potential for damage to the stone during the pouring of the concrete, however, it is still probably more feasible to develop such a structure as a veneered one, with the concrete wall cast first and the stone anchored to it later.

6.3 CONSTRUCTION WITHOUT MORTAR

Low retaining structures for site construction are sometimes built of stone without mortar. These may be of a general rubble form in random pattern (see Fig. 2.3) or achieved with flat stones in coursed layers, as shown in Fig. 6.4. Lack of need to relate to supported elements or other construction in general makes the general dimensional stability of the construction much less critical.

The unmortared construction may actually be a better choice in many cases, since it can adjust in minor ways as it settles into place without development of the cracking that often occurs in similar structures of brick or CMUs. The structure must nevertheless keep some level of stability to remain functional, and the same amount of care is required in fitting the units as with mortared stone work.

Another advantage of the stone structure without mortar is its ability to drain water through the structure, preventing the buildup behind the retaining structure, which is a concern for solid walls in the same situation.

Although we do not now consider using stone construction without mortar for buildings, this was the basic form of the construction in ancient times.

7

MISCELLANEOUS MASONRY CONSTRUCTION

Masonry construction has been, and still can be, produced with many different materials in various forms. The principal topic of this book is structural masonry as presently used in the United States, which is represented primarily in the treatments in Chapters 4 and 5. Nevertheless, there are other form of masonry in use for special purposes or simply because of enduring attachments to historic procedures and symbols. This chapter presents some of the other forms of masonry construction that have endured.

7.1 ADOBE

Adobe is a term that refers to a masonry unit (a sun-dried unburned brick), the construction process in which it is used, and the building produced with the process. It is a very ancient form of construction, probably the antecedent of all other forms of brick construction.

Adobe bricks are simply produced with soil materials that somewhat emulate a good concrete mix. There must be a binder (cement–clay) and preferably a graded aggregate (for adobe a mixture of silt and fine-to-coarse sand). This just happens to be the normal constituency of the surface soils in most temperate, arid regions, exactly where abode construction thrives. So, if you live in Arizona and want to make adobe bricks, just go out in your backyard and scoop up some dirt, mix a little water with it, and make bricks.

To make a wall, you lay out a row of bricks, fill the cracks between them with the same mud you used to make the bricks, spread a layer of mud on top of them, and lay another row of bricks on top of the mud, continuing the process until the wall is as high as you want it. Normal time required for the hand labor (including siestas and festivals) will ensure that lower courses dry well and that shrinkage in the slowly drying mass is steadily and incrementally accumulative, resulting in a very stable structure.

Play this scene in the century of your choice, and you have the basic adobe process, accessible to persons with very little developed technology, economic system, or scientific knowledge. Just the earth, the sun, some water, and their hands—an enduring building culture, if there ever was one.

Figure 7.1 is a reproduction of a page from the third edition (1941) of *Architectural Graphic Standards,* presenting a modern version of the ages-old process with a few high-tech touches, such as steel industrial windows, flashing, anchor bolts, and strengthening with reinforced concrete members. Also indicated are the basic techniques of protecting the soft, moisture-susceptible adobe with an exterior coating of stucco (cement plaster) and reinforcing the stucco with good old chicken wire. Pragmatic references to "cheap" and "good" construction are also used for communication with the practical builder.

It is interesting to note that the latest (eighth) edition of *Architectural Graphic Standards* (Ref. 11) has three pages of adobe construction.

A variation of adobe brick construction is the use of rammed earth, in which the same basic material (sand–silt–clay) is used in a manner similar to sitecast concrete. Forms are erected, and shallow layers of the soil are tamped (rammed) into the forms to form a dense mass. A variation on this is the use of soil–cement, consisting of the addition of a small quantity of portland cement to the soil mixture. Soil–cement construction may also be used to produce foundations for ordinary adobe buildings.

Adobe is fundamentally brick construction, and all of the ideas and tricks for improving and strengthening masonry in general can be used, and in fact were probably largely learned originally in building with adobe. Use of pilasters, lintels, general strengthening of openings, corners, and wall intersections, and bonding of multiple wythes apply equally with adobe construction.

Bricks for adobe construction are ordinarily made for use in single-wythe walls. The standard brick is therefore quite wide, usually 10 to 12 in. It is also usually made as large as possible, to reduce the number required and speed the construction. The unit size therefore more closely relates to that of current CMUs, and a popular form of CMU construction today uses 4-in.- or 6-in.-high units with a bulging side form (called slump block) that consciously imitates ancient construction with adobe. (See Fig. 7.2.)

Because of its enduring popularity, modern building codes usually give some acknowledgment to adobe and provide some criteria for its design and construction. (See Sec. 2407(i)6 in the 1988 *UBC*.)

7.2 GLASS BLOCK

Large units of glass in block form were produced many years ago for use in partitions and exterior walls, producing light-transmitting but generally vision-blocking dividers. These became an architectural icon of the modern styles in the 1930s and 1940s. As that style became dated, their use receded, but a recent resurgence has occurred.

Although the typically heavy glass units are usually quite strong, their use is generally limited to nonstructural applications. The highly durable material, however, makes them quite practical for exterior exposure, and a principal usage is for glazing. Large walls can be built up by integrating a light steel frame with the glass units, the glass serving primarily as infill but also helping to brace the slender frame elements.

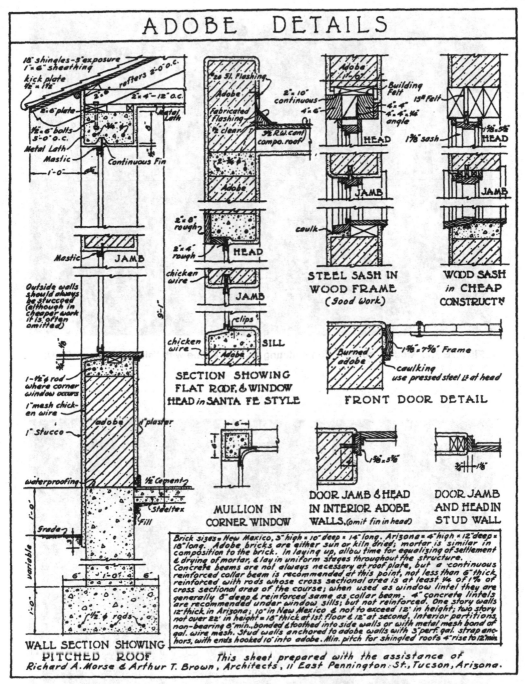

FIGURE 7.1. Details for adobe construction. Reproduced from *Architectural Graphic Standards*, 3rd edition, 1941, with permission of the publishers, John Wiley & Sons.

FIGURE 7.2. Modern construction imitating historic adobe with "slump" form of CMUs.

FIGURE 7.3. Use of glazed tile, imitating elaborate carved stone.

7.3 CLAY TILE

Man-made masonry units are generally formed of fired clay or cast concrete. Small units are generally solid, while large units are made with significant voids. In earlier times, fired clay was used to produce large, voided units, called tiles or tile blocks. Although now largely displaced by CMUs, the clay units were forerunners of present CMUs, and much of the detail of present CMU construction was originally developed with clay tile units.

A special form of clay tile unit is referred to as architectural terra-cotta. These consist of units with glazed surfaces, intended for exposure to view. They were developed as economical imitations of cut stone and widely used for elaborate cornices and other decorative features on buildings in the form of architecture that prevailed in the nineteenth century and early part of the twentieth century. Since this coincided with a major use of masonry, in general, the use of clay tile was perpetuated until other materials were developed to imitate it. (See Fig. 7.3.)

Fired clay can quite easily attain strength superior to most cast concrete, so clay tile units were characteristically quite thin and light despite the relatively large size of units. The large units, however, resulted in a rather inflexible modular system, so finished wall dimensions were often achieved by patching in with bricks at ends, tops, and openings.

8

USE OF STRUCTURAL MASONRY
FOR BUILDINGS

Masonry was originally developed primarily as a structural material and utilized extensively for exterior, load-bearing walls. Thus, much of the historical heritage of architecture consists of buildings whose exterior surfaces strongly express the use of masonry materials. This has resulted in a considerable public affection for masonry and an association of the materials of masonry with qualities of endurance and permanence in construction. Now, what appears to be masonry in new construction is largely not structural masonry; in fact it is frequently not masonry at all. This chapter presents some considerations for the use of masonry for buildings, with emphasis on concerns other than structural ones. Much of this material is generally applicable to masonry in general, although the concentration is on forms of masonry that serve structural functions.

8.1 REGIONAL CONSIDERATIONS

In these times, architectural styles tend to be global in usage. Nevertheless, usage of materials and basic forms of construction are often quite regional in character. This is especially true of masonry, for some of the following reasons.

Use of Regionally Produced Materials

Many factors work to influence the use of local products for general construction use. For masonry a particular concern is for the bulk use and weight of the materials, making transportation over great distances mostly impractical. Thus brick may be favored over CMUs in a particular region simply because there are a lot of good bricks locally available and market competition is strong.

Climate

Local climate conditions may also favor or preclude certain materials or forms of construction. Walls without insulation may be reasonable in very mild climates, but are currently unacceptable in cold climates. Thermal range, amount of rainfall, degree of freezing, wind storm conditions, and risk of earthquakes are all regional concerns, and will affect choices of materials and forms of construction.

Influence of Codes

Building codes and local industry standards largely control design and construction of masonry structures in a given location. These sources steadily tend to be more national in scope and form, but regional preferences and interests can exert influence. Vested interests of local craft unions, masonry contractors, masonry product manufacturers and suppliers, and masonry industry agencies often work to influence local codes and ordinances.

Continuity of Local Experience and Customs

There is nothing like demonstrated success to encourage people to continue the use of methods of building. When a major amount of local construction of a particular type has been successfully used for many years and many local landmarks were produced with it, it tends to get firmly entrenched. It takes more than a shift in styles or claims of better performance to dislodge well-established methods.

Regional differences tend to get superficially obliterated in this time of rapid mass communication. Downtown Anchorage, Alaska, looks a lot like downtown Tucson, Arizona. In fact, there are many good practical reasons for regional differences in use of building methods and materials. If anything, it is likely that not enough at-

tention is given to logical concern for regional issues, and illogical architectural styles and details are imported in the name of conformance with the latest design trends. This is a lesson that could be learned from history, but the learning ever goes on.

Some simple lessons regarding regional use of masonry are the following:

1. Unreinforced masonry should not be used in areas with histories of severe wind storms or earthquakes. Actually, most building codes exclude it.

2. Control joints and other details relating to relief of stresses due to thermal expansion and contraction are much more critical in cold climates, where the seasonal range of outdoor temperature is greater.

3. Care must be exercised in development of the details for installation of reinforcement in masonry exposed to the weather in regions with severe freezing or high rainfall, since the possibility for rusting is greatly increased.

4. Exterior walls must be insulated in cold climates. This is generally easier to achieve with nonstructural masonry. Form and details must be developed with regard for architectural concerns, as well as structural and other physical-response concerns.

5. Investigate the availability of specific materials and any special construction methods in the region of a proposed building: the smaller the project, the greater this concern.

6. Study the local, legally applicable building codes for all requirements relating to masonry construction. Codes are legal ordinances, enacted by some government body, and frequently lag behind the very latest national standards.

8.2 CHOICE OF CONSTRUCTION

Choice of construction materials and methods for any building is subject to many influences. Preferences of the building owners or designers must be considered, but other sources of influence may be hard to ignore. Without consideration for rank or value, the following are some sources of concern:

1. *Codes and Standards.* Enforceable codes and currently prevailing industry standards will tend to strongly govern the details of design and construction. Choices should be made from forms of construction that are truly viable alternatives, not ideas with little likelihood of use.
2. *Cost and General Feasibility.* Many factors will influence this, and all considerations should be included for any specific case. Exact costs fluctuate, but relative costs change less often over time. For exposed work some higher cost may be justified to obtain a desired appearance. But for strictly workhorse, structural applications, money talks. (See the general discussion of building economics in Sec. 9.8.)
3. *Knowledge of Alternatives.* For any situation in building design there are typically some alternative means for achieving design requirements in terms of different forms of construction or use of materials. This requires some research, experience, and access to reliable information.

8.3 DETAILS OF CONSTRUCTION

Designs for buildings are eventually communicated in the form of construction plans, details, and specifications. For structures, the development of appropriate details is as much a part of the design work as are the structural computations.

There are many sources for structural details and many variations for similar situations. Designers and builders often have personal preferences and styles. Industry standards sometimes exist, but are also forthcoming from more than one source in some cases. The latest information from the most respected sources should be given preference.

Many details are regional in character, especially those dealing with the building's exterior surfaces. Care must be exercised in using details from a particular building or from a single regional source. Local codes, climate conditions, and favored forms of construction must be acknowledged.

Finally, and most importantly, details are rapidly dated, as materials and products change, codes are revised, relative cost fluctuates, and recent failures disprove designer's or manufacturer's claims.

We have used many details for illustration in this book, but none should be considered to have authoritative status. For general reference we recommend References 7, 9, 10, and 11. All of these publications, however, are dated.

8.4 ENHANCEMENTS

Masonry walls are sometimes used in an unadorned state. However, when they serve as building walls, especially as exterior walls, they frequently receive some additional treatment—that is, some form of enhancement of the basic masonry structure.

Applied Finishes

A common form of addition is that of an applied surface finish. This occurs most often with a wall of CMUs, which is the general system of choice when the masonry serves essentially utilitarian tasks, such as a structural backup. Finishes of just about any kind can be applied, both for interior and exterior situations. The main concern for the masonry is the form of attachment of the finish. If attachment requires the incorporation of preset elements (anchor bolts, threaded inserts, etc.), these must be clearly described as part of the mason's work, and any relation to the joint layouts, joint sizes, interference with other elements, such as joint reinforcement, or other possible concerns must be studied in development of the general construction details.

A vast array of hardware exists for the association of all forms of masonry with all imaginable forms of finish treatments. Use of these materials is not generally a specific concern for the structural designer of the masonry work. However, where usage of particular forms of attachment requires any special provisions in the masonry, it behooves the structural designer to be aware of the considerations required.

Thermal Insulation

Solid masonry walls are not generally effective as barriers to thermal flow. There are two different situations to consider in this regard: cold weather and hot weather. When it is hot outside, it can get very uncomfortable inside, even though the temperature difference between a reasonably comfortable indoor condition (75°F?) and a hot outdoor condition (100°F+) is not great (25°F±). Thermal flow through the enclosure is thus less critical for summer cooling, compared to the effects of humidity and air infiltration.

Winter conditions in cold climates, however, represent a major temperature differential between indoors and outdoors, as much as 100°F or more. In this situation, the thermal flow barrier effect of the enclosure becomes very important. In cold climates, therefore, it is imperative that effective insulation be added to exterior walls.

Various priority systems exist for use in developing masonry with enhanced resistance to thermal flow. (See Fig. 5.3.) It is also possible to develop resistance by means generally applicable to any form of wall construction, namely by addition of fiberglass batts or sheets of foamed plastic, which can be incorporated in stud spaces or surface-adhered to masonry or concrete.

The need for insulation, the exact amount required, the cost effectiveness of various forms, the special accommodation required by the masonry construction, and the relationships to the general architectural development of the wall make a complex systems design problem, of which the structural design of the masonry is only one factor. As in all situations in the design of building structures, the designer must realize the broader context in which the structural design decisions must be considered.

One method for improving the insulative effect of the masonry is to make a thermal break in the construction. This can be achieved almost totally with veneered construction that is separated from its backup support by an airspace. A partial effect of this kind is also achieved with an unfilled cavity in brick construction or the unfilled void spaces in hollow-block construction. The airspace itself represents an interruption of the thermal flow; however, it can also be filled with insulative material for an increased effect.

Because of its typical density and mass, a masonry wall also generally represents a major potential storage element of heat, which may be at least as important as its thermal barrier function in some situations. This factor must be considered in a total evaluation of the thermal influence of the construction.

Thermal Mass and Radiation

As mentioned in the preceding paragraph, the mass of a masonry wall functions as

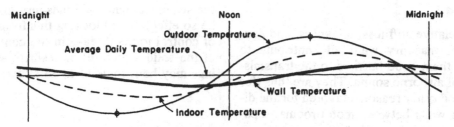

FIGURE 8.1. Daily temperature swings and the thermal inertial effect.

a major storage element for heat. In weather conditions where a major swing of daily temperatures occurs, the inertial effect of this mass can be significant. Figure 8.1 shows the general form of temperature swings in a 24-hour day, from a low in the early morning to a high in the mid-afternoon on a clear day. Interior spaces in enclosures tend to follow this form of the outdoor swing, with a minor lag caused by the enclosure. If the enclosure is a good thermal barrier, the indoor swing may be considerably less in total magnitude from high to low.

In addition to its barrier effects, the building contributes an inertial effect in the form of the heat stored in its mass. This is a three-way exchange between the outdoor air and sun, the indoor air and occupants, and the thermal mass of the building as a radiant energy device. The building is cooled at night and releases its cooling effect by absorbing heat from the interior at midday. Then it is heated during the day and reverses the effect in the evening to early morning hours.

To achieve the effect just described, the building thermal mass must be largely exposed to outdoor conditions, a situation generally true if exterior walls are of massive masonry. The effect is mostly useful, however, only in relatively mild climates with high daily temperature swings (arid, southwestern United States, for example). Insulating the walls from the interior somewhat diminishes this effect, although the total effect of the combination must be studied for complete effectiveness.

In a very cold climate, where outdoor temperatures remain low over long periods—day and night—the thermal mass works better if it is kept indoors, that is, at indoor temperature. This is best achieved by putting the insulation on the *outside* of the walls. Brick veneers with airspaces and actual exterior insulations are effective in this situation.

Water Penetration

The building enclosure must block the penetration of water from precipitation and melting ice and snow. Good solid masonry with tight mortar joints and no major cracks will block direct flow of water, but most masonry materials are water-absorptive to some degree. Water absorption can be lowered by addition of materials to concrete units or by coating or glazing of bricks, but at some point the good adherence of mortar becomes a problem.

Wall surfaces may be coated with materials to inhibit moisture penetration. This may also be done to reduce efflorescence, help with curing of the mortar, and generally protect the wall finish.

Mostly, however, water tends to penetrate at joints, both those between masonry units and those between the masonry and other elements of the construction, such as window frames. Good detailing and the correct use of flashing and sealants are imperative to ensure that neither water nor air will penetrate the joints.

Sound

The relative stiffness, thickness, and density of masonry walls all contribute to make them good barriers to the transmission of airborne sound. They are thus (for this and other reasons) favored for the dividing walls between motel rooms, separate apartments, and other spaces where privacy is a concern.

However, as in all situations, sound can travel through very small openings, such as cracks around doors, edges of ceilings, or around electrical or other equipment built into walls. It can also flank the wall entirely, going out one window and into an adjacent one, for example. Nevertheless, a good barrier begins with the basic wall construction, and masonry is usually at an advantage in this regard.

Noise can also be transmitted through the structure, and solid masonry may not be so effective in blocking this form. Use of control joints (breaks in the continuity of the solid structure) or resilient surfacings may help some of these situations.

Fire

Continuous masonry walls usually represent good barriers to the spread of fire and are often chosen where major barriers are required. As with airborne sound, fire can leak through cracks or flank a wall, so attention to these details of the construction is also required. Although the basic masonry materials may be resistant to surface spread of fires, applied surfacings can drastically change this situation, reducing some of the fire-resistive potential of the construction.

9

GENERAL CONCERNS FOR BUILDING STRUCTURES

This chapter contains some discussions of general issues relating to design of building structures. These concerns have mostly not been addressed in the presentations in earlier chapters, but require some general consideration when dealing with whole building design situations. General application of these materials is illustrated in the design examples in Chapter 10.

9.1 INTRODUCTION

Materials, methods, and details of building construction vary considerably on a regional basis. There are many factors that affect this situation, including the real effects of response to climate and the availability of construction materials. Even in a single region, differences occur between individual buildings, based on individual styles of architectural design and personal techniques of builders. Nevertheless, at any given time there are usually a few predominant, popular methods of construction that are employed for most buildings of a given type and size. The construction methods and details shown here are reasonable, but in no way are they intended to illustrate a singular, superior style of building.

9.2 DEAD LOADS

Dead load consists of the weight of the materials of which the building is constructed, such as walls, partitions, columns, framing, floors, roofs, and ceilings. In the design of a beam, the dead load must include an allowance for the weight of the beam itself. Table 9.1, which lists the weights of many construction materials, may be used in the computation of dead loads. Dead loads are due to gravity and they result in downward vertical forces.

Dead load is generally a permanent load, once the building construction is completed, unless frequent remodeling or rearrangement of the construction occurs. Because of this permanent, long-time, character, the dead load requires certain considerations in design, such as the following:

1. It is always included in design loading combinations, except for investigations of singular effects, such as deflections due to only live load.
2. Its long-time character has some special effects causing sag and requiring reduction of design stresses in wood structures, producing creep effects in concrete structures, and so on.
3. It contributes some unique responses, such as the stabilizing effects that resist uplift and overturn due to wind forces.

Dead load is of particular concern with masonry structures because the construc-

Table 9.1 WEIGHTS OF BUILDING CONSTRUCTION

	lb/ft^2	kN/m^2
Roofs		
3-ply ready roofing (roll, composition)	1	0.05
3-ply felt and gravel	5.5	0.26
5-ply felt and gravel	6.5	0.31
Shingles		
Wood	2	0.10
Asphalt	2–3	0.10–0.15
Clay tile	9–12	0.43–0.58
Concrete tile	8–12	0.38–0.58
Slate, $\frac{1}{4}$ in.	10	0.48
Fiber glass	2–3	0.10–0.15
Aluminum	1	0.05
Steel	2	0.10
Insulation		
Fiber glass batts	0.5	0.025
Rigid foam plastic	1.5	0.075
Foamed concrete, mineral aggregate	2.5/in.	0.0047/mm
Wood rafters		
2 × 6 at 24 in.	1.0	0.05
2 × 8 at 24 in.	1.4	0.07
2 × 10 at 24 in.	1.7	0.08
2 × 12 at 24 in.	2.1	0.10
Steel deck, painted		
22 ga	1.6	0.08
20 ga	2.0	0.10
18 ga	2.6	0.13
Skylight		
Glass with steel frame	6–10	0.29–0.48
Plastic with aluminum frame	3–6	0.15–0.29
Plywood or softwood board sheathing	3.0/in.	0.0057/mm
Ceilings		
Suspended steel channels	1	0.05
Lath		
Steel mesh	0.5	0.025
Gypsum board, $\frac{1}{2}$ in.	2	0.10
Fiber tile	1	0.05
Drywall, gypsum board, $\frac{1}{2}$ in.	2.5	0.12
Plaster		
Gypsum, acoustic	5	0.24
Cement	8.5	0.41
Suspended lighting and air distribution Systems, average	3	0.15

Table 9.1 *(Continued)*

	lb/ft²	kN/m²
Floors		
Hardwood, ½ in.	2.5	0.12
Vinyl tile, ⅛ in.	1.5	0.07
Asphalt mastic	12/in.	0.023/mm
Ceramic tile		
¾ in.	10	0.48
Thin set	5	0.24
Fiberboard underlay, ⅝ in.	3	0.15
Carpet and pad, average	3	0.15
Timber deck	2.5/in.	0.0047/mm
Steel deck, stone concrete fill, average	35–40	1.68–1.92
Concrete deck, stone aggregate	12.5/in.	0.024/mm
Wood joists		
2 × 8 at 16 in.	2.1	0.10
2 × 10 at 16 in.	2.6	0.13
2 × 12 at 16 in.	3.2	0.16
Lightweight concrete fill	8.0/in.	0.015/mm
Walls		
2 × 4 studs at 16 in., average	2	0.10
Steel studs at 16 in., average	4	0.20
Lath, plaster; see Ceilings		
Gypsum drywall, ⅝ in. single	2.5	0.12
Stucco, ⅞ in., or wire and paper or felt	10	0.48
Windows, average, glazing + frame		
Small pane, single glazing, wood or metal frame	5	0.24
Large pane, single glazing, wood or metal frame	8	0.38
Increase for double glazing	2–3	0.10–0.15
Curtain walls, manufactured units	10–15	0.48–0.72
Brick veneer		
4 in., mortar joints	40	1.92
½ in., mastic	10	0.48
Concrete block		
Lightweight, unreinforced—4 in.	20	0.96
6 in.	25	1.20
8 in.	30	1.44
Heavy, reinforced, grouted—6 in.	45	2.15
8 in.	60	2.87
12 in.	85	4.07

tion tends to be quite heavy in comparison with structures that are essentially frameworks with lightweight infill. Settlements of foundations may be more critical and may cause distress for the stiff, brittle masonry. On the other hand, the dead load, as a permanent load, may be an advantage as a stabilizing or anchoring source. Holding the rest of the building down against the effects of wind uplift or overturn may be a significant contribution. For seismic effects, however, the increased mass represents an increased inertial force, adding to the lateral load.

As mentioned elsewhere in this book, the solid, heavy mass of the masonry can be an asset as a thermal inertia element, helping to sustain the interior temperature conditions through swings in the outdoor temperature.

9.3 BUILDING CODE REQUIREMENTS

Structural design of buildings is most directly controlled by building codes, which are the general basis for the granting of building permits—the legal permission required for construction. Building codes (and the permit-granting process) are administered by some unit of government: city, county, or state. Most building codes, however, are based on some model code, of which there are three widely used in the United States:

1. The *Uniform Building Code, (UBC,* Ref. 1), which is widely used in the West because it has the most complete data for seismic design.

2. *The BOCA Basic National Building Code,* used widely in the East and Midwest.

3. *The Standard Building Code,* used in the Southeast.

These model codes are more similar than different, and are in turn largely derived from the same basic data and standard reference sources, including many industry standards. In the several model codes and many city, county, and state codes, however, there are some items that reflect particular regional concerns.

With respect to control of structures, all codes have materials (all essentially the same) that relate to the following issues:

1. *Minimum Required Live Loads.* This is addressed in Sec. 9.4; all codes have tables similar to those shown in Tables 9.2 and 9.3, which are reproduced from the *UBC.*

2. *Wind Loads.* These are highly regional in character with respect to concern for local windstorm conditions. Model codes provide data with variability on the basis of geographic zones.

3. *Seismic(Earthquake)Effects.* These are also regional with predominant concerns in the western states. This data, including recommended investigations, is subject to quite frequent

modification, as the area of study responds to ongoing research and experience.

4. *Load Duration.* Loads or design stresses are often modified on the basis of the time span of the load, varying from the life of the structure for dead load to a fraction of a second for a wind gust or a single major seismic shock. Safety factors are frequently adjusted on this basis. Some applications are illustrated in the work in the design examples in this part.

5. *Load Combinations.* These were formerly mostly left to the discretion of designers, but are now quite commonly stipulated in codes, mostly because of the increasing use of ultimate strength design and the use of factored loads.

6. *Design Data for Types of Structures.* These deal with basic materials (wood, steel, concrete, masonry, etc.), specific structures (towers, balconies, pole structures, etc.), and special problems (foundations, retaining walls, stairs, etc.) Industry-wide standards and common practices are generally recognized, but local codes may reflect particular local experience or attitudes. Minimal structural safety is the general basis, and some specified limits may result in questionably adequate performances (bouncy floors, cracked plaster, etc.)

7. *Fire Resistance.* For the structure, there are two basic concerns, both of which produce limits for the construction. The first concern is for structural collapse or significant structural loss. The second concern is for containment of the fire to control its spread. These concerns produce limits on the choice of materials (e.g., combustible or non-

combustible) and some details of the construction (cover on reinforcement in concrete, fire insulation for steel beams, etc.)

The work in the design examples in this part is based largely on criteria from the *UBC*.

9.4 LIVE LOADS

Live loads technically include all the nonpermanent loadings that can occur, in addition to the dead loads. However, the term as commonly used usually refers only to the vertical gravity loadings on roof and floor surfaces. These loads occur in combination with the dead loads, but are generally random in character and must be dealt with as potential contributors to various loading combinations, as discussed in Sec. 9.3.

Roof Loads

In addition to the dead loads they support, roofs are designed for a uniformly distributed live load that includes snow accumulation and the general loadings that occur during construction and maintenance of the roof. Snow loads are based on local snowfalls and are specified by local building codes.

Table 9.2 gives the minimum roof live-load requirements specified by the 1988 edition of the *UBC*. Note the adjustments for roof slope and for the total area of roof surface supported by a structural element. The latter accounts for the increase in probability of the lack of total surface loading as the size of the surface area increases.

Roof surfaces must also be designed for wind pressure, for which the magnitude and manner of application are specified by local building codes based on local wind histories. For very light roof construction,

a critical problem is sometimes that of the upward (suction) effect of the wind, which may exceed the dead load and result in a net upward lifting force.

Although the term *flat roof* is often used, there is generally no such thing; all roofs must be designed for some water drainage. The minimum required pitch is usually $\frac{1}{4}$ in./ft, or a slope of approximately 1:50. With roof surfaces that are this close to flat, a potential problem is that of *ponding*, a phenomenon in which the weight of water on the surface causes deflection of the supporting structure, which in turn allows for more water accumulation (in a pond), causing more deflection, and so on, resulting in an accelerated collapse condition.

Floor Loads

The live load on a floor represents the probable effects created by the occupancy. It includes the weights of human occupants, furniture, equipment, stored materials, and so on. All building codes provide minimum live loads to be used in the design of buildings for various occupancies. Since there is a lack of uniformity among different codes in specifying live loads, the local code should always be used. Table 9.3 contains values for floor live loads as given by the *UBC*.

Although expressed as uniform loads, code-required values are usually established large enough to account for ordinary concentrations that occur. For offices, parking garages, and some other occupancies, codes often require the consideration of a specified concentrated load as well as the distributed loading. Where buildings are to contain heavy machinery, stored materials, or other contents of unusual weight, these must be provided for individually in the design of the structure.

When structural framing members support large areas, most codes allow some reduction in the total live load to be used

Table 9.2   MINIMUM ROOF LIVE LOADS

ROOF SLOPE	METHOD 1			METHOD 2		
	TRIBUTARY LOADED AREA IN SQUARE FEET FOR ANY STRUCTURAL MEMBER			UNIFORM LOAD[2]	RATE OF REDUC- TION r (Percent)	MAXIMUM REDUC- TION R (Percent)
	0 to 200	201 to 600	Over 600			
1. Flat or rise less than 4 inches per foot. Arch or dome with rise less than one eighth of span	20	16	12	20	08	40
2. Rise 4 inches per foot to less than 12 inches per foot. Arch or dome with rise one eighth of span to less than three eighths of span	16	14	12	16	06	25
3. Rise 12 inches per foot and greater. Arch or dome with rise three eighths of span or greater	12	12	12	12		
4. Awnings except cloth covered[3]	5	5	5	5	No Reductions Permitted	
5. Greenhouses, lath houses and agricultural buildings[4]	10	10	10	10		

[1]Where snow loads occur, the roof structure shall be designed for such loads as determined by the building official. See Section 2305 (d). For special purpose roofs, see Section 2305 (e).

[2]See Section 2306 for live load reductions. The rate of reduction r in Section 2306 Formula (6-1) shall be as indicated in the table. The maximum reduction R shall not exceed the value indicated in the table.

[3]As defined in Section 4506.

[4]See Section 2305 (e) for concentrated load requirements for greenhouse roof members.

Source: Adapted from the *Uniform Building Code*, 1988 ed. (Ref. 1), copyright © 1988, with the permission of the publishers, the International Conference of Building Officials.

for design. These reductions, in the case of roof loads, are incorporated into the data in Table 9.2. The following is the method given in the *UBC* for determining the reduction permitted for beams, trusses, or columns that support large floor areas.

Except for floors in places of assembly (theaters, etc.), and except for live loads greater than 100 psf [4.79 kN/m²], the design live load on a member may be reduced in accordance with the formula

$$R = 0.08(A - 150)$$

$$[R = 0.86(A - 14)]$$

The reduction shall not exceed 40% for horizontal members or for vertical members receiving load from one level only, 60% for other vertical members, nor R as determined by the formula

$$R = 23.1\left(1 + \frac{D}{L}\right)$$

Table 9.3   MINIMUM FLOOR LOADS

USE OR OCCUPANCY		UNIFORM LOAD[1]	CONCENTRATED LOAD
CATEGORY	DESCRIPTION		
1. Access floor systems	Office use	50	2000[2]
	Computer use	100	2000[2]
2. Armories		150	0
3. Assembly areas[3] and auditoriums and balconies therewith	Fixed seating areas	50	0
	Movable seating and other areas	100	0
	Stage areas and enclosed platforms	125	0
4. Cornices, marquees and residential balconies		60	0
5. Exit facilities[4]		100	0[5]
6. Garages	General storage and/or repair	100	[6]
	Private or pleasure-type motor vehicle storage	50	[6]
7. Hospitals	Wards and rooms	40	1000[2]
8. Libraries	Reading rooms	60	1000[2]
	Stack rooms	125	1500[2]
9. Manufacturing	Light	75	2000[2]
	Heavy	125	3000[2]
10. Offices		50	2000[2]
11. Printing plants	Press rooms	150	2500[2]
	Composing and linotype rooms	100	2000[2]
12. Residential[7]		40	0[5]
13. Rest rooms[8]			
14. Reviewing stands, grandstands and bleachers		100	0
15. Roof deck	Same as area served or for the type of occupancy accommodated		
16. Schools	Classrooms	40	1000[2]
17. Sidewalks and driveways	Public access	250	[6]
18. Storage	Light	125	
	Heavy	250	
19. Stores	Retail	75	2000[2]
	Wholesale	100	3000[2]

[1]See Section 2306 for live load reductions.
[2]See Section 2304 (c), first paragraph, for area of load application.
[3]Assembly areas include such occupancies as dance halls, drill rooms, gymnasiums, playgrounds, plazas, terraces and similar occupancies which are generally accessible to the public.
[4]Exit facilities shall include such uses as corridors serving an occupant load of 10 or more persons, exterior exit balconies, stairways, fire escapes and similar uses.
[5]Individual stair treads shall be designed to support a 300-pound concentrated load placed in a position which would cause maximum stress. Stair stringers may be designed for the uniform load set forth in the table.
[6]See Section 2304(c), second paragraph, for concentrated loads.
[7]Residential occupancies include private dwellings, apartments and hotel guest rooms.
[8]Rest room loads shall be not less than the load for the occupancy with which they are associated, but need not exceed 50 pounds per square foot.

Source: Adapted from the *Uniform Building Code,* 1988 ed. (Ref. 1), copyright © 1988, with the permission of the publishers, the International Conference of Building Officials.

In these formulas

R = reduction, in percent

A = area of floor supported by a member

D = unit dead load/ft^2 of supported area

L = unit live load/ft^2 of supported area

In office buildings and certain other building types, partitions may not be permanently fixed in location but may be erected or moved from one position to another in accordance with the requirements of the occupants. In order to provide for this flexibility, it is customary to require an allowance of 15–20 psf [0.72–0.96 kN/m^2], which is usually added to other dead loads.

9.5 LATERAL LOADS

As used in building design, the term *lateral load* is usually applied to the effects of wind and earthquakes, as they induce horizontal forces on stationary structures. From experience and research, design criteria and methods in this area are continuously refined, with recommended practices being presented through the various model building codes, such as the *UBC*.

Space limitations do not permit a complete discussion of the topic of lateral loads and design for their resistance. The following discussion summarizes some of the criteria for design in the latest edition of the *UBC*. Examples of application of these criteria are given in the chapters that follow containing examples of building structural design. For a more extensive discussion the reader is referred to *Simplified Building Design for Wind and Earthquake Forces* (Ref. 15).

Wind

Where wind is a major local problem, local codes are usually more extensive with regard to design requirements for wind. However, many codes still contain relatively simple criteria for wind design. One of the most up-to-date standards for wind design is contained in the *American National Standard Minimum Design Loads for Buildings and Other Structures*, ANSI A58.1-1982 (Ref. 2), published by the American National Standards Institute in 1982.

Complete design for wind effects on buildings includes a large number of both architectural and structural concerns. The following is a discussion of some of the requirements for wind as taken from the 1988 edition of the *UBC*, which is in general conformance with the material presented in the ANSI standard just mentioned.

Basic Wind Speed. This is the maximum wind speed (or velocity) to be used for specific locations. It is based on recorded wind histories and adjusted for some statistical likelihood of occurrence. For the continental United States the wind speeds are taken from *UBC*, Fig. No. 4. As a reference point, the speeds are those recorded at the standard measuring position of 10 m (approximately 33 ft) above the ground surface.

Exposure. This refers to the conditions of the terrain surrounding the building site. The ANSI standard describes four conditions (A, B, C, and D), although the *UBC* uses only two (B and C). Condition C refers to sites surrounded for a distance of one-half mile or more by flat, open terrain. Condition B has buildings, forests, or ground surface irregularities 20 ft or more in height covering at least 20% of the area for a distance of 1 mile or more around the site.

Wind Stagnation Pressure (q$_s$). This is the basic reference equivalent static pressure based on the critical local wind speed. It is given in *UBC* Table No. 23-F and is based on the following formula as given in the ANSI standard:

$$q_s = 0.00256V^2$$

Example. For a wind speed of 100 mph,

$$q_s = 0.00256V^2 = 0.00256(100)^2$$
$$= 25.6 \text{ psf [1.23 kPa]}$$

which is rounded off to 26 psf in the *UBC* table.

Design Wind Pressure. This is the equivalent static pressure to be applied normal to the exterior surfaces of the building and is determined from the formula

$$p = C_e C_q q_s I$$

(*UBC* Formula 11-1, Sec. 2311), in which

p = design wind pressure in psf

C_e = combined height, exposure, and gust factor coefficient as given in *UBC* Table No. 23-G

C_q = pressure coefficient for the structure or portion of structure under consideration as given in *UBC* Table No. 23-H

q_s = wind stagnation pressure at 30 ft given in *UBC* Table No. 23-F

I = importance factor

The importance factor is 1.15 for facilities considered to be essential for public health and safety (such as hospitals and government buildings) and buildings with 300 or more occupants. For all other buildings the factor is 1.0.

The design wind pressure may be positive (inward) or negative (outward, suction) on any given surface. Both the sign and the value for the pressure are given in the *UBC* table. Individual building surfaces, or parts thereof, must be designed for these pressures.

Design Methods. Two methods are described in the Code for the application of the design wind pressures in the design of structures. For design of individual elements particular values are given in *UBC* Table 23-H for the C_q coefficient to be used in determining p. For the primary bracing system the C_q values and their use is to be as follows:

Method 1 (Normal Force Method). In this method wind pressures are assumed to act simultaneously normal to all exterior surfaces. This method is required to be used for gabled rigid frames and may be used for any structure.

Method 2 (Projected Area Method). In this method the total wind effect on the building is considered to be a combination of a single inward (positive) horizontal pressure acting on a vertical surface consisting of the projected building profile and an outward (negative, upward) pressure acting on the full projected area of the building in plan. This method may be used for any structure less than 200 ft in height, except for gabled rigid frames. This is the method generally employed by building codes in the past.

Uplift. Uplift may occur as a general effect, involving the entire roof or even the whole building. It may also occur as a local phenomenon such as that generated by the overturning moment on a single shear wall. In general, use of either design method will account for uplift concerns.

Overturning Moment. Most codes require that the ratio of the dead load resisting moment (called the restoring moment, stabilizing moment, etc.) to the overturning moment be 1.5 or greater. When this is not the case, uplift effects must be resisted by anchorage capable of developing the excess overturning moment. Overturning may be a critical problem for the whole building, as in the case of relatively tall and slender tower structures. For buildings braced by individual shear walls, trussed bents, and rigid-frame bents, overturning is investigated for the individual bracing units. Method 2 is usually used for this investigation, except for very tall buildings and gabled rigid frames.

Drift.   Drift refers to the horizontal deflection of the structure due to lateral loads. Code criteria for drift are usually limited to requirements for the drift of a single story (horizontal movement of one level with respect to the next above or below). The *UBC* does not provide limits for wind drift. Other standards give various recommendations, a common one being a limit of story drift to 0.005 times the story height (which is the *UBC* limit for seismic drift). For masonry structures wind drift is sometimes limited to 0.0025 times the story height. As in other situations involving structural deformations, effects on the building construction must be considered; thus the detailing of curtain walls or interior partitions may affect limits on drift.

Combined Loads.   Although wind effects are investigated as isolated phenomena, the actions of the structure must be considered simultaneously with other phenomena. The requirements for load combinations are given by most codes, although common sense will indicate the critical combinations in most cases. With the increasing use of load factors the combinations are further modified by applying different factors for the various types of loading, thus permitting individual control based on the reliability of data and investigation procedures and the relative significance to safety of the different load sources and effects. Required load combinations are described in Sec. 2303 of the *UBC*.

Special Problems.   The general design criteria given in most codes are applicable to ordinary buildings. More thorough investigation is recommended (and sometimes required) for special circumstances such as the following:

Tall Buildings. These are critical with regard to their height dimension as well as the overall size and number of occupants inferred. Local wind speeds and unusual wind phenomena at upper elevations must be considered.

Flexible Structures. These may be affected in a variety of ways, including vibration or flutter as well as the simple magnitude of movements.

Unusual Shapes. Open structures, structures with large overhangs or other projections, and any building with a complex shape should be carefully studied for the special wind effects that may occur. Wind-tunnel testing may be advised or even required by some codes.

Use of code criteria for various ordinary buildings is illustrated in the design examples in Chapter 10.

Earthquakes

During an earthquake a building is shaken up and down and back and forth. The back-and-forth (horizontal) movements are typically more violent and tend to produce major unstabilizing effects on buildings; thus structural design for earthquakes is mostly done in terms of considerations for horizontal (called lateral) forces. The lateral forces are actually generated by the weight of the building—

or, more specifically, by the mass of the building that represents both an inertial resistance to movement and the source for kinetic energy once the building is actually in motion. In the simplified procedures of the equivalent static force method, the building structure is considered to be loaded by a set of horizontal forces consisting of some fraction of the building weight. An analogy would be to visualize the building as being rotated vertically 90° to form a cantilever beam, with the ground as the fixed end and with a load consisting of the building weight.

In general, design for the horizontal force effects of earthquakes is quite similar to design for the horizontal force effects of wind. Indeed, the same basic types of lateral bracing (shear walls, trussed bents, rigid frames, etc.) are used to resist both force effects. There are indeed some significant differences, but in the main a system of bracing that is developed for wind bracing will most likely serve reasonably well for earthquake resistance as well.

Because of its considerably more complex criteria and procedures, we have chosen not to illustrate the design for earthquake effects in the examples in this part. Nevertheless, the development of elements and systems for the lateral bracing of the buildings in the design examples here is quite applicable in general to situations where earthquakes are a predominant concern. For structural investigation, the principal difference is in the determination of the loads and their distribution in the building. Another major difference is in the true dynamic effects, critical wind force being usually represented by a single, major, one-direction punch from a gust, while earthquakes represent rapid back-and-forth, reversing-direction actions. However, once the dynamic effects are translated into equivalent static forces, design concerns for the bracing systems are very similar, involving considerations for shear, overturning, horizontal sliding, and so on.

For a detailed explanation of earthquake effects and illustrations of the investigation by the equivalent static force method the reader is referred to *Simplified Building Design for Wind and Earthquake Forces* (Ref. 15).

Masonry structures have several properties that make them potentially very vulnerable to damage from earthquakes. Major concerns are the following:

Weight. Lateral seismic forces are generated by the impelled mass of the building, and masonry construction is generally quite heavy.

Stiffness. Flexible structures dissipate some of the dynamic effects of seismic shock through their deformations; masonry is typically quite rigid and generally nondeforming.

Brittleness. The tension-weak masonry materials are subject to sudden failure by brittle cracking.

Good design may lessen some of these factors, but the unreinforced masonry structure remains generally undesirable in terms of seismic resistance. Extensive use of steel reinforcement is the primary means for improving seismic resistance.

9.6 STRUCTURAL PLANNING

Planning a structure requires the ability to perform two major tasks. The first is the logical arranging of the structure itself, regarding its geometric form, its actual dimensions and proportions, and the ordering of the elements for basic stability and reasonable interaction. All of these issues must be faced, whether the building is simple or complex, small or large, of ordinary construction or totally unique. Spanning beams must be supported and have depths adequate for the spans; thrusts of arches must be resolved; columns above should be centered over columns below; and so on.

The second major task in structural planning is the development of the relationships between the structure and the building in general. The building plan must be "seen" as a structural plan. The two may not be quite the same, but they must fit together. "Seeing" the structural plan (or possibly alternative plans) inherent in a particular architectural plan is a major task for designers of building structures.

Hopefully, architectural planning and structural planning are done interactively, not one after the other. The more the architect knows about structural problems and the structural designer (if another person) knows about architectural problems, the more likely it is possible that an interactive design development may occur.

Although each building offers a unique situation if all of the variables are considered, the majority of building design problems are highly repetitious. The problems usually have many alternative solutions, each with its own set of pluses and minuses in terms of various points of comparison. Choice of the final design involves the comparative evaluation of known alternatives and the eventual selection of one.

The word *selection* may seem to imply that all the possible solutions are known in advance, not allowing for the possibility of a new solution. The more common the problem, the more this may be virtually true. However, the continual advance of science and technology and the fertile imagination of designers make new solutions an ever-present possibility, even for the most common problems. When the problem is truly a new one in terms of a new building use, a scale jump, or a new performance situation, there is a real need for innovation. Usually, however, when new solutions to old problems are presented, their merits must be compared to established previous solutions in order to justify them. In its broadest context the selection process includes the consideration of all possible alternatives: those well known, those new and unproven, and those only imagined.

For masonry structures a planning consideration that must be noted is the dimensions established by details of the masonry units and the general type of construction. Although bricks come in a wide variety of sizes, once a particular brick is chosen its specific dimensions set some required modules for the construction. Bricks can be cut, but this is generally not feasible with regard to their vertical dimension. Thus vertical dimensions of the masonry construction should relate to the repeating module of the brick height plus the mortar joint.

Construction with CMUs represents an even more restrictive situation, since it is not generally feasible to cut the units to a trimmed size. Many special sizes and shapes are available for various situations, but the construction in general should be carefully planned to utilize the materials in a logical and reasonable way.

Structural masonry walls are generally quite thick, and their true dimensions must be carefully allowed for in architectural planning. The general forms of walls should be established early in the building design process, with thicknesses established that allow for unit sizes, multiwythe buildups, and any cavities in the construction.

Structural masonry walls often support other structural systems for roofs or floors. Modular planning units of the masonry must be coordinated with dimensional units of the supported systems, such as the center-to-center spacing of joists and rafters.

9.7 SYSTEMS INTEGRATION

Good structural design requires integration of the structure into the whole physical system of the building. It is necessary to realize the potential influences of structural design decisions on the general architectural design and on the development of the systems for power, lighting, thermal control, ventilation, water supply, waste handling, vertical transportation, firefighting, and so on. The most popular structural systems have become so in many cases largely because of their ability to accommodate the other subsystems of the building and to facilitate popular architectural forms and details.

Masonry structures present somewhat greater difficulty in accommodating the incorporation of various building service elements, such as wiring, piping, wall outlets and switches, recessed lighting, and other built-in items. Hollow construction with frames of wood or steel provide some very convenient spaces and easier construction procedures for such elements. Various techniques can be used to deal with this problem, but it does indeed add some additional planning work for the masonry structure.

9.8 ECONOMICS

Dealing with dollar cost is a very difficult, but necessary, part of structural design. For the structure itself, the bottom-line cost is the delivered cost of the finished structure, usually measured in units of dollars per square foot of the building. For individual components, such as a single wall, units may be used in other forms. The individual cost factors or components, such as cost of materials, labor, transportation, installation, testing, and inspection, must be aggregated to produce a single unit cost for the entire structure.

Designing for control of the cost of the structure is only one aspect of the design problem, however. The more meaningful cost is that for the entire building construction. It is possible that certain cost-saving efforts applied to the structure may result in increases of cost for other parts of the construction. A common example is that of the floor structure for multistory buildings. Efficiency of floor beams occurs with the generous provision of beam depth in proportion to the span. However, adding inches to beam depths with the unchanging need for dimensions required for floor and ceiling construction and installation of ducts and lighting elements means increasing the floor-to-floor distance and the overall height of the building. The resulting increases in cost for the added building skin, interior walls, elevators, piping, ducts, stairs, and so on, may well offset the small savings in cost of the beams. The really effective cost-reducing structure is often one that produces major savings of nonstructural costs, in some cases at the expense of less structural efficiency.

Real costs can only be determined by those who deliver the completed construction. Estimates of cost are most reliable in the form of actual offers or bids for the construction work. The farther the cost estimator is from the actual requirement to deliver the goods, the more speculative the estimate. Designers, unless they are in the actual employ of the builder, must base any cost estimates on educated guesswork deriving from some comparison with similar work recently done in the same region. This kind of guessing must be adjusted for the most recent developments in terms of the local markets, competitiveness of builders and suppliers, and the general state of the economy. Then the four best guesses are placed in a hat and one is drawn out.

Serious cost estimating requires a lot of

training and experience and an ongoing source of reliable, timely information. For major projects various sources are available, in the form of publications or computer-accessible data bases.

The following are some general rules for efforts that can be made in the structural design work in order to have an overall, general cost-saving attitude.

1. Reduction of material volume is usually a means of reducing cost. However, unit prices for different grades must be noted. Higher-quality materials may be proportionally more expensive than the higher stress values they represent; more volume of cheaper material may be less expensive.

2. Use of standard, commonly stocked products is usually a cost savings, since special sizes or shapes may be premium priced.

3. Reduction in the complexity of systems is usually a cost savings. Simplicity in purchasing, handling, managing of inventory, and so on, will be reflected in lower bids as builders anticipate simpler tasks. Use of the fewest number of different grades of materials, sizes of fasteners, and other such variables is as important as the fewest number of different parts. This is especially true for any assemblage done on the building site; large inventories may not be a problem in a factory, but usually are on the site.

4. Cost reduction is usually achieved when materials, products, and construction methods are highly familiar to local builders and construction workers. If real alternatives exist, choice of the "usual" one is the best course.

5. Do not guess at cost factors; use real experience, either yours or others. Costs vary locally, by job size, and over time. Keep up to date with cost information.

6. In many situations, labor cost is greater than material cost. A dozen or so bricks must be used to provide the same masonry volume as a single CMU, representing considerable more work for the mason and a generally significant increase in cost. Savings in labor, especially at the building site, is usually much more significant than savings in materials for reducing overall building cost.

7. For buildings of an investment nature, time is money. Speed of construction may be a major advantage. However, getting the structure up fast is not a true advantage unless the other aspects of the construction can take advantage of the time gained. Steel frames often go up quickly, only to stand around and rust while the rest of the work catches up.

10

BUILDING STRUCTURES: DESIGN EXAMPLES

This chapter presents examples of the design of structural systems for buildings. The buildings used for the examples have been chosen to create a range of situations in order to be able to demonstrate the use of various structural components. What is of primary concern in this chapter is the illustration of the design process for whole systems and the consideration of the many factors that influence design decisions. Many of these factors are not structural in nature, but nevertheless have major effects on the decisions regarding the final form and details of the construction.

Masonry is hardly ever used to create entire buildings. It is necessary, therefore, to show other forms of construction for roofs and floors, and sometimes for foundations, columns, and nonstructural walls. Although the designs of these other elements are discussed only briefly here, they have been shown in sufficient detail to illustrate their relations to the masonry structures. Commonly used elements have been used for the examples, although regional differences may well result in different choices for materials, systems, and details of the construction.

Design of individual elements of structural masonry is largely based on materials presented in earlier chapters. In order to conserve space, references are made to computations demonstrated in the earlier chapters, so few computations are shown here for some examples.

10.1 BUILDING 1

Building 1 consists of a box-shaped, one-story structure that is intended for commercial occupancy. Figure 10.1 shows a scheme for the construction that uses structural exterior walls of CMUs and a clear-span roof structure with open-web steel joists and a formed steel deck. We assume the following for design data:

Roof live load = 20 psf (reducible)
Design wind pressure = 20 psf (assumed, *UBC* method 2)

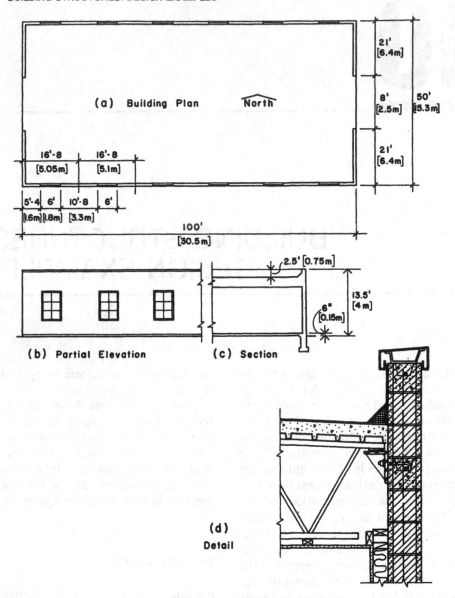

FIGURE 10.1. Building 1.

CMUs: medium-weight units, grade N, ASTM C90, f'_m = 1350 psi, mortar type S

The general profile of the building is shown in Fig. 10.1c, which indicates a low-sloping roof, a flat ceiling, and a short parapet at the exterior walls. The general form of the construction is shown in the wall section in Fig. 10.1d. Design for lat-

eral loads is discussed in Sec. 10.3. We first consider the design for gravity loads.

10.2 BUILDING 1: DESIGN FOR GRAVITY LOADS

The masonry walls serve as bearing walls to support the roof construction. For the dead weight of the roof construction, we determine the following (see Table 9.1):

Three-ply felt and gravel roofing	5.5 psf
Foamed concrete insulating fill, 4 in.	10.0
Formed sheet steel deck, 20 ga.	2.0
Open web joists, from suppliers catalog	12.0
Ceiling: wood nailers and blocking,	1.0
gypsum drywall	2.5
Lighting, HVAC, etc.	3.0
Total roof dead load	36.0 psf

Ignoring for the moment the openings in the wall, we determine the uniformly distributed load on the wall per foot of wall length as follows:

1. Roof dead load = (25 ft)(36 psf) = 900 lb/ft
2. Roof live load = (25 ft)(20 psf) = 500 lb/ft (not reduced)
3. Estimating the wall dead weight at an average of 60 psf per square foot of wall surface, we find the total weight at the bottom of the wall:

$$\text{Wall dead load} = (13.5 \text{ ft})(60)$$
$$= 810 \text{ lb/ft}$$

At the top of the foundation wall, therefore, the total design load is 900 + 500 + 810 = 2210 lb/ft. If we use a nominal 8-in.-thick wall (actually 7.5 in., typically) and assume a block with a 50% void, the unit bearing stress in compression at the base of the wall is

$$f_a = \frac{P}{A} = \frac{2210}{(7.5)(12)(0.50)} = 49 \text{ psi}$$

The investigation and design of a bearing wall is discussed in Sec. 5.5. The roof joists (actually light trusses) represent concentrated loads on the top of the wall. Although the average bearing stress in the wall may be low, the load concentration may be considerable if the trusses are widely spaced. To provide for this, even in so-called unreinforced construction, it is common to place horizontal reinforcement in the top course of the CMUs, thus creating a continuous beam of sorts on top of the wall. It is also possible, of course, to create individual pilaster columns at the locations of the trusses, although this is usually not necessary unless the trusses are very long in span, are quite widely spaced (8 ft or more on center), or the roof construction is excessively heavy. (See construction for alternative 1 for building 2.)

The wall loading just determined is for the north and south walls, since the roof trusses span the narrow dimension of the building. The loading will thus be less on the east and west walls. However, the north and south walls have large openings for windows, so the actual bearing at the bottom of the walls is not uniform along the whole wall length. There will be some concentration of loading in the solid portions of the wall at the edges of the window openings, due to the spanning of the upper wall over the openings. The details of the construction for achieving the 6-ft span over the openings and resisting the load concentration at the edge of the openings will depend partly on the form of the masonry.

If the masonry is reinforced—that is, block voids are vertically aligned and voids are filled with reinforcement and concrete at regular intervals—a lintel over the opening will be developed as a horizontally reinforced beam, and edges of the openings will be developed as reinforced columns. If unreinforced construction is used, it is most common to use a steel lintel over the opening, although a sitecast or precast concrete lintel could

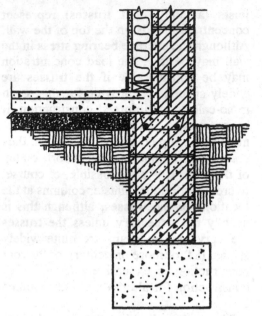

FIGURE 10.2. Building 1: masonry foundation wall.

also be integrated into the block arrangement.

The section in Fig. 10.2 shows the use of a short foundation wall and a footing for support of the wall. If a concrete masonry wall is used, as shown in Fig. 10.2, it will ordinarily be constructed as shown, with all block voids filled with concrete and some horizontal reinforcement in the top of the wall. This wall will generally serve to distribute the gravity loads in a uniform manner to the footing, with an assumed loading condition as shown in Fig. 10.3. The reinforcement in the top is thus useful for resistance of the tension occurring as the wall spans between openings.

The short foundation wall could also be made of sitecast concrete, as shown in Fig. 10.4b. Or, as is common when severe freezing of the ground is not a problem, the foundation wall and footing could be combined into a single element called a grade beam, as shown in Fig. 10.4a. Other factors, such as soil conditions, roof drainage systems, building underground utilities, or structures for site plantings may affect the decision as to the most desired form of construction.

If a separate footing is used, as shown in Fig. 10.2 or 10.4b, it could be selected from tables of predesigned foundations in various references (for example, Ref. 6,

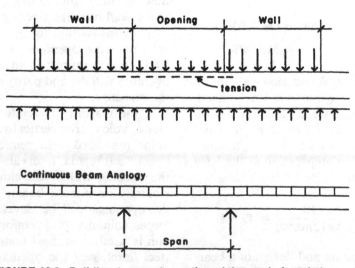

FIGURE 10.3. Building 1: spanning action of the grade foundation walls.

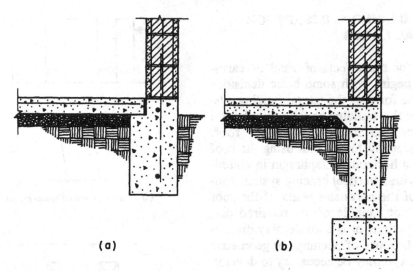

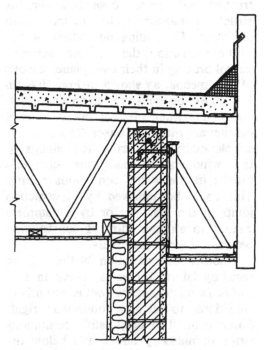

FIGURE 10.4. Building 1: sitecast concrete options for the foundations.

11, 13 and 14). With the light roof construction in this example, a quite narrow footing will suffice, and it would probably be constructed without transverse reinforcement, since it would project only a few inches beyond the edges of the supported wall.

A consideration for the wall design is the manner in which the roof loading is transmitted to the wall. If the parapet wall is used and the trusses are supported as shown in Fig. 10.1d, the roof gravity load is not placed on the centroidal axis of the wall and a bending is induced in the wall due to its eccentricity. The wall must be designed for the combined bending and compression, as discussed in Sec. A.5. The procedures for this depend on the form of the wall construction, as reinforced or unreinforced.

Additionally, of course, the walls must be designed for lateral effects of wind or earthquakes, which is discussed in the next section. In the end, the wall design must satisfy the various possible critical loading combinations, as discussed in Sec. 9.5.

A modification of the construction that generally eliminates the bending caused by gravity loading is shown in Fig. 10.5. Here the roof trusses bear on top of the wall, and a short cantilevered roof edge and soffit are developed.

FIGURE 10.5. Building 1: alternative detail for the roof edge.

10.3   BUILDING 1: DESIGN FOR LATERAL LOADS

Design for the effects of wind or earthquakes begins with some basic decisions about the form of the lateral bracing system. For this building consisting only of a perimeter wall structure and a flat roof, the simplest system is that using the roof deck as a horizontal diaphragm in combination with a vertical bracing system consisting of the perimeter walls. If the roof deck is not capable of the required diaphragm actions or is considerably discontinuous because of openings or geometric variations, it may be necessary to develop some other form of horizontal bracing, such as a horizontal truss system. However, in this case, for the size of the building, the formed sheet steel deck is most likely adequate for the required diaphragm tasks.

The perimeter bracing could consist of units of moment-resistive rigid frames or trussed bents. In this case, however, the structural masonry walls are more than sufficient. Depending on details of the wall construction, the walls may serve as lateral bracing in their own planes by one of two means, as shown in Fig. 10.6. In Fig. 10.6*a* the walls are shown as consisting of individual piers acting as vertical cantilevers with fixed bases. This action is developed by considering the openings for the windows to constitute complete breaks in the wall's continuous nature. This could be achieved by construction joints in the masonry or by a complete change to another form of construction between the masonry piers.

Another form of action for the walls in resisting lateral forces is shown in Fig. 10.6*b*. In this case, the wall is essentially considered to act as a continuous, rigid-frame bent. The deep, stiff, continuous strips of masonry above and below the windows are considered to be nonflexing elements, and the piers of masonry be-

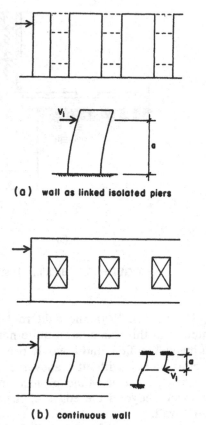

(a)   wall as linked isolated piers

(b)   continuous wall

FIGURE 10.6. Building 1: functioning of the masonry shear walls: (*a*) as isolated, linked piers; (*b*) as a continuous pierced wall.

tween the windows acts as columns, fixed at their tops and bottoms. If the masonry is built as reinforced construction, and the portions above and below the windows are indeed quite deep, this is a reasonable assumption for the wall action.

For the individual pier action shown in Fig. 10.6*a*, the single wall units will be designed as individual, freestanding shear walls. This involves considerations for the following:

1. *Horizontal Shear in the Wall.* This unit stress is determined by simply dividing the horizontal force by the horizontal cross section of the wall. An allowable stress (or ultimate de-

sign strength) is established, and the materials, details, and any necessary reinforcing of the wall is determined.

2. *Bending (Cantilever Beam Action) of the Wall.* Depending on the wall form and construction, the bending may be resisted by the entire wall or developed essentially by the two opposed wall ends acting like flanges of an I-beam. For reinforced masonry with concrete blocks, the latter is usually assumed, with the two end "columns" acting in compression and tension.

3. *Anchorage of the Wall for Overturning Effect (Rotation at the Base).* The basis for this investigation is shown in Fig. 10.7. Appropriate safety factors are used to ensure that the total resistance exceeds the overturn effect. See discussion in Sec. 5.8.

4. *Development of the Force Transfers.* This involves the considerations of the construction details to achieve the transfer of force from the roof diaphragm to the piers and from the piers to their supports. Attachment of the roof framing to the wall will achieve the roof-to-wall transfer. Ordinarily, the doweling of the vertical foundation by reinforcement (see Fig. 10.2 or 10.4*b*) will achieve the necessary base tiedown effect (*T* in Fig. 10.7)

For any form of construction there are usually some minimum requirements that establish a base level of capability once the form of construction is chosen. For structural masonry, the usual code requirements for the masonry units, mortar, and some details of the construction will establish this minimal construction. In many applications, for buildings of modest size, the structural capacity of this minimal construction will not be exceeded. Thus nothing extra need be done to develop the required structural actions. Such is the case for this building. If properly detailed and built to code standards, either reinforced or unreinforced masonry could achieve the building shown in Fig. 10.1, with no significant "extras" required.

Masonry shear walls are ordinarily quite heavy, so that many have sufficient dead weight to counteract the overturning effects for shear wall action (Fig. 10.7). Thus the tie-down provided by foundation dowels is a bonus, but not really required for the lateral force resistance.

Figure 10.8 shows the basis for consideration of the horizontal force effects of wind on building 1. Uplift on the roof must also be considered, but the effects illustrated in Fig. 10.8 are the usual basis for design of the roof deck as a horizontal diaphragm and the determination of the lateral loads on the perimeter bracing.

The walls must also act as spanning elements in resisting the direct wind pressures on their surfaces. As shown in Fig. 10.8*a*, the walls span from floor to roof, acting in one of the two ways shown. If the construction is as shown in Fig. 10.1*d*, and the wall is continuous past the roof edge, it will act as a beam with an overhanging end (Case 1). If the construction is as shown in Fig. 10.4, the parapet is developed in conjunction with the roof construction, and the wall spans simply between the floor and roof levels (Case 2).

In either of the cases shown in Fig.

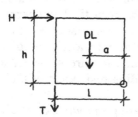

FIGURE 10.7. Basis for stability analysis of the cantilever shear wall.

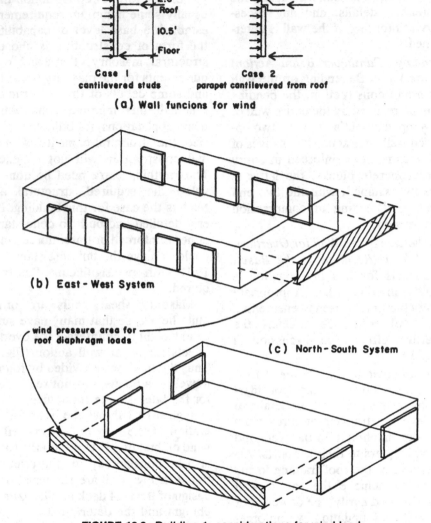

FIGURE 10.8. Building 1: considerations for wind load.

10.8a, some of the wind load on the wall goes directly into the edge of the floor and is not delivered to the edge of the roof diaphragm. Thus the wind load to be used for the investigation of the roof and the shear walls is that shown in Fig. 10.8b (for east or west direction wind) or Fig. 10.8c (for north or south direction wind).

The spanning action of the walls (Fig. 10.8a) results in bending in the walls, which must be combined with the actions required by the gravity loads. For tall walls and/or very high wind pressures, this may be a critical concern. Bending due to wind may add to any bending caused by gravity loads (as discussed in Sec. A.5) and require some enhancement of the bending resistance of the construction.

Considering the wind on the north and south walls and assuming a wall action as shown for case 2 in Fig. 10.8a, the wind load delivered to the roof edge is

$$(20 \text{ psf}) \left(\frac{10.5}{2}\right) + (20 \text{ psf})(2.5) = 155 \text{ lb/ft}$$

In resisting this load, the roof functions as a spanning member supported by the shear walls at the east and west ends of the building. The investigation of this 100-ft-span simple beam is shown in Fig. 10.9. The roof diaphragm must sustain the resulting shear (in the steel deck) and bending (as edge compression or tension in the edge framing members). The end reaction forces of 7.75 kips each become the lateral loads to the end walls and are distributed to the individual elements of the walls in proportion to their relative stiffnesses. Determined thusly, the loadings for the individual wall piers on all sides of the building are shown in Fig. 10.10. This assumes the individual pier form of action (Fig. 10.6a) and the pattern of walls shown in Fig. 10.1d and e.

For many of the available systems and components used for buildings such as building 1, rated capacities are established by codes or by the manufacturers of products. "Design" thus becomes largely a matter of matching load requirements, as determined by a loading investigation, to capacities of available elements. This is done most easily with a computer-aided process, but can also be done quite readily

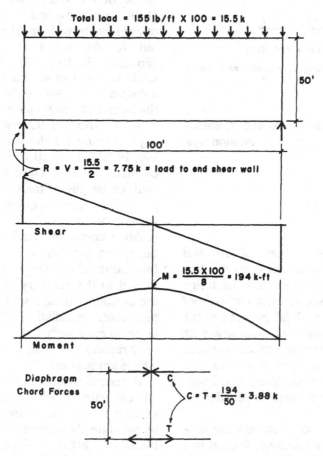

FIGURE 10.9. Building 1: functioning of the roof diaphragm.

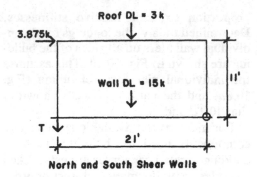

North and South Shear Walls

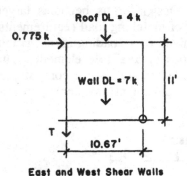

East and West Shear Walls

FIGURE 10.10. Building 1: stability analysis of the shear walls.

with available handbooks and manufacturer's brochures for very common systems.

10.4 BUILDING 2: ALTERNATIVE 1

Figure 10.11 shows the general form and some details for a small one-story building for a library. A major feature of the design is the use of exposed elements of the roof and wall structures. Solid portions of the exterior walls consist of masonry with CMUs with the exterior surfaces exposed to view. The clear-span roof structure consists of large precast, prestressed concrete T-sections, with their undersides exposed to view.

The T-shaped roof elements are supported by masonry columns, formed in a monolithic mass with the masonry walls.

The roof structure extends on all sides to form a 4-ft cantilevered overhang.

In mild climates it may be possible to use the masonry walls with no additional construction, exposing their surfaces to view on both the interior and exterior. For other than very utilitarian uses, however (parking garages, warehouses, etc.), this is usually not desirable. Use of interior masonry surfaces complicates attachments to the wall surface and the installation of wiring and wall outlets. The solid masonry is also not very insulative, although special forms of insulation can be built into some types of construction.

For various purposes, an addition to the construction frequently consists of the attachment of a furred-out space on the inside of the wall, with wood or metal strips (like studs) and an applied surfacing of gypsum drywall or other materials. Details for the furred-out construction are shown in Fig. 10.11. The furred-out space could be quite small, but the details show a deeper space with batts of insulation in the furred-out void space.

Two additional alternatives for providing insulation for the walls are shown in Fig. 10.12. Figure 10.12*a* shows a layered construction adhered to the inside of the wall, consisting of thick sheets of foamed plastic with gypsum drywall applied to the surface of the insulation. Depending on building codes and manufacturer's directions, the insulation and finish may be fully adhered or may be mechanically attached to the masonry surface; materials are available to achieve either form of attachment. Installation of wiring is not quite so easy with this system, but it is a very rapidly installed, economical system where appropriate.

A special system is that shown in Fig. 10.12*b*, with a layered installation on the outside of the walls. This is generally used to simulate the appearance of stucco, although the actual final finish is an acrylic plastic. One advantage of this system is

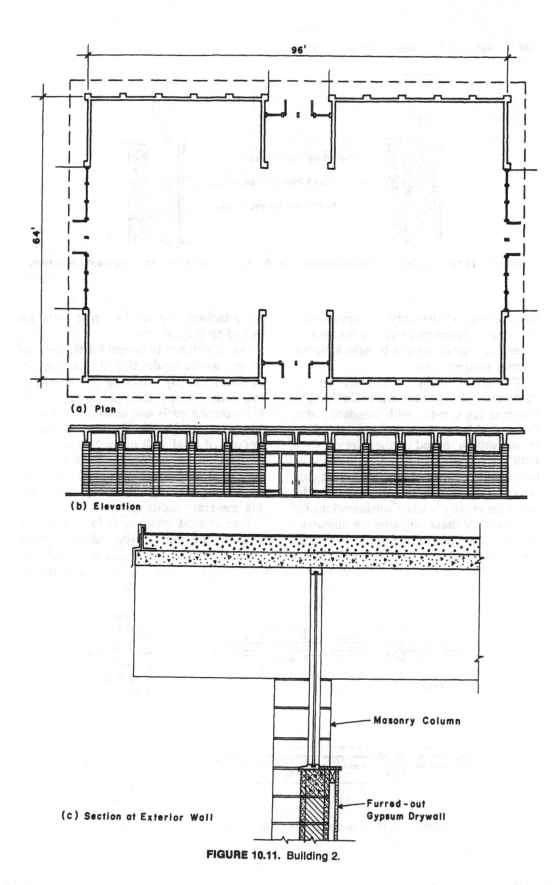

(a) Plan

(b) Elevation

(c) Section at Exterior Wall

Masonry Column

Furred-out
Gypsum Drywall

96'

64'

FIGURE 10.11. Building 2.

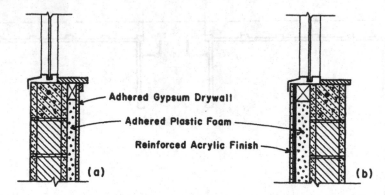

FIGURE 10.12. Building 2: alternative details for the exterior walls with adhered foam insulation.

that the mass of masonry is maintained at a warmer temperature and can function as a thermal inertial source to help maintain interior temperatures.

The details in Fig. 10.11 show the use of reinforced masonry with CMUs. The columns are created with standard units, forming 16-in.-square pilasters, which become partly exposed on the interior, even with the furred-out construction. Despite the considerable dead load of the roof construction, these columns will be quite adequate for gravity loads. Considerations for the design of these columns are discussed in Sec. 5.10. The bearing of the T-units on top of the columns would be achieved with steel elements cast into the T-units

and attached to a steel bearing plate anchored to the column.

For resistance to lateral loads, the roof system would be developed as a horizontal diaphragm by attaching the T-units to each other along their adjoining edges. The exterior walls and columns will function as shear walls, with a reasonable amount of total wall in each direction, if all the walls can serve this function. Ordinary minimum construction for the reinforced masonry will probably suffice for this one-story building.

One critical concern is for the lateral shear in the tops of the columns, especially if seismic load effects are critical. Closely spaced ties should be used to rein-

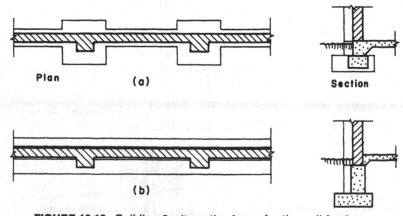

FIGURE 10.13. Building 2: alternative forms for the wall footings.

force the extended tops above the top of the walls.

Because of the heavy roof, there will be a much greater gravity load at the columns, compared with the remainder of the exterior walls. This may make it necessary to consider the form of bearing-footing layout in Fig. 10.13a, with square footings under the columns and a minimum-width footing under the walls between the column footings. However, the columns are quite closely spaced, and it may be possible to consider the use of a continuous-width footing, as shown in Fig. 10.13b.

The continuous-width footing would be simpler in construction and possibly less costly, even though some additional concrete and reinforcement might be required. An additional factor in this decision has to do with the depth of the footings below grade. If frost protection or a drop to better bearing material is a concern, a grade wall as shown in Fig. 10.13b might be used and could be reinforced to serve as a grade beam to spread the column loads.

10.5 BUILDING 2: ALTERNATIVE 2

An alternative construction to building 2 is shown in Fig. 10.14. The masonry wall and column construction are preserved, but the roof structure consists of a site-cast, two-way spanning, concrete joist system, called waffle construction. The waffle can be formed with a variety of patented or custom-designed systems; that shown here is developed with a common process using sheet-steel-formed "pans" (like upside-down dishpans) that are placed in crisscrossed rows, with the spaces between pans forming the ribs of the waffle.

The waffle shown uses a pan width of 30 in. with a 6-in.-wide rib, forming a 3-ft plan grid. This modular dimension does not mesh well with the usual 8-in. plan module of the CMUs, so some adjustments must be made to connect the concrete and masonry systems. One place that this can be achieved is at the exterior walls, where a widened rib is used over the columns and windows. The other plan adjustments are made in the places where windows and doors break the module of the CMU system.

Figure 10.15 shows a partial reflected plan of the waffle system (like looking down at a mirror on the floor). Because of the high concentration of shear and bending at the interior column, a solid portion is created at this location by simply leaving out a group of the forming pans. The form of this solid element shown in Fig. 10.15 is the usual one, with a recommended side dimension of approximately one third the waffle span.

There are fewer columns in the exterior walls with this scheme. Both the gravity and lateral loads will thus be greater on each column, and the column size must be carefully considered for the lateral shear and combined loadings. The interior column is shown as round on the plans, which would probably be achieved with a concrete column. It is possible, however, to use a masonry column or a steel column at this location. Loading is primarily only gravity here, so the major concern is for the simple axial compression and the development of the connection at the waffle support.

Because of the wider spacing and higher loads for the exterior columns, the foundation scheme here is likely to be that shown in Fig. 10.13a.

10.6 BUILDING 3

Building 3 is a three-story office building, designed for speculative rental. As with buildings 1 and 2, there are many alternatives for the construction of such a build-

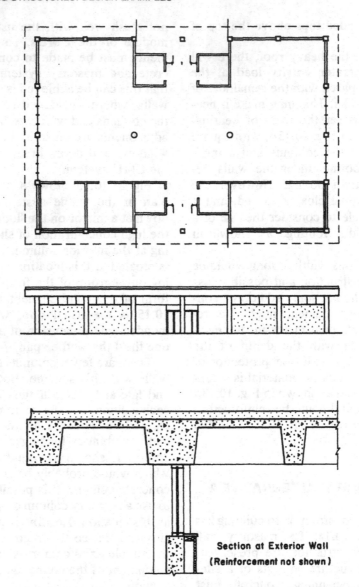

FIGURE 10.14. Building 2: alternative structure with waffle construction. Drawings show building plan and elevation, wall and roof section detail, and partial reflected plan of waffle.

ing, although in any given place at any given time the basic construction of such buildings will most likely vary little from a limited set of choices. We will show here some of the work for the design of two alternatives for the construction using structural masonry walls.

General Considerations for the Building Design

Figure 10.16 presents a plan of the upper floor and a full building section for building 3. We assume that a fundamental requirement for the building is the provision

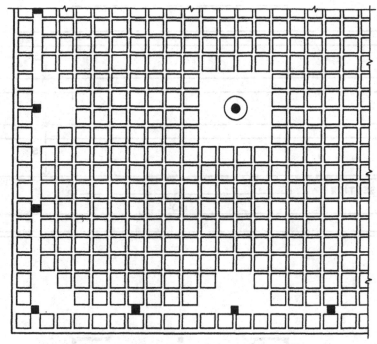

FIGURE 10.15. Building 2: reflected plan of the waffle system.

of a significant amount of exterior window surface and the avoidance of long expanses of unbroken solid wall surface. Another assumption is that the building is freestanding on the site so that all sides have a clear view. Most designers would also prefer that the space available for rental be as free as possible of permanent construction, permitting maximum flexibility in rearrangement for successive tenants. The latter translates to a general desire to eliminate permanent structural elements (columns or bearing walls) from the rental space, preferring them at the building perimeter and the location of permanent plan elements such as stairs, elevators, restrooms, and duct shafts.

The following will be assumed as criteria for the design work:

Building code: 1988 edition of *Uniform Building Code* (Ref. 1)

Live Loads

Roof: Table 9.2 (*UBC* Table 23-C)

Floors: Table 9.3 (*UBC* Table 23-A)

Office areas: 50 psf [2.39 kPa]

Corridor and lobby: 100 psf [4.79 kPa]

Partitions: 20 psf (*UBC* minimum per Sec. 2304) [0.96 kPa]

Wind: map speed, 80 mph [129 km/h]; exposure B

Assumed construction loads

Floor finish: 5 psf [0.24 kPa]

Ceilings, lights, ducts: 15 psf [0.72 kPa]

Walls (average surface weight)

Interior, permanent: 10 psf [0.48 kPa]

Exterior curtain wall: 15 psf [0.72 kPa]

Structural Alternatives

The plan as shown, with 30-ft-square bays and a general open interior, is an ideal arrangement for a beam and column system

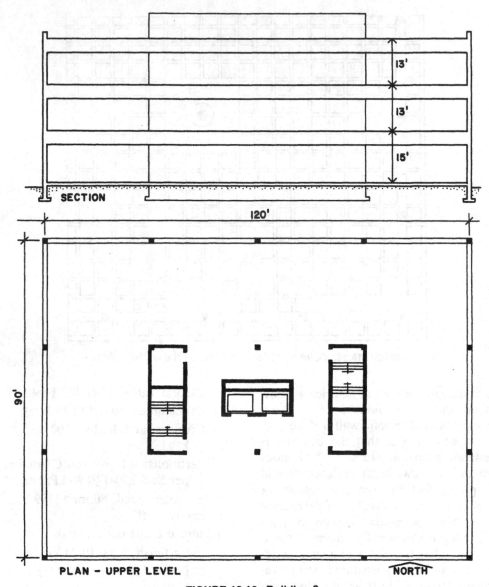

FIGURE 10.16. Building 3.

in either steel or reinforced concrete. Other types of systems may be made more effective if some modifications of the basic plans are made. These changes may affect the planning of the building core, the plan dimensions for the column locations, the articulation of the exterior wall, or the vertical distances between the levels of the building.

The general form and basic type of the structural system must relate to both the gravity and lateral force problems. Considerations for gravity require the development of the horizontal spanning systems for the roof and floors and the arrangement of the vertical elements (walls and columns) that provide support for the spanning structure. Vertical elements should be stacked, thus requiring coordinating the plans of the various levels.

The most common choices for the lateral bracing system would be the following (see Fig. 10.17):

1. *Core Shear Wall System* (Fig. 10.17*a*). This consists of using solid walls to produce a very rigid central core. The rest of the structure leans on this rigid interior portion, and the roof and floor constructions outside the core, as well as the exterior walls, are free for concerns for lateral forces as far as the structure as a whole is concerned.

2. *Truss-Braced Core*. This is similar in nature to the shear-wall-braced core, and the planning considerations would be essentially similar.

The solid walls would be replaced by bays of trussed framing (in vertical bents) using various possible patterns for the truss elements.

3. *Peripheral Shear Walls* (Fig. 10.17*b*). This in essence makes the building into a tubelike structure. Because doors and windows must pierce the exterior, the peripheral shear walls usually consist of linked sets of individual walls (sometimes called piers).

4. *Mixed Exterior and Interior Shear Walls*. This is essentially a combination of the core and peripheral systems.

5. *Full Rigid-Frame System* (Fig. 10.17*c*). This is produced by using the vertical planes of columns and beams in each direction as a series

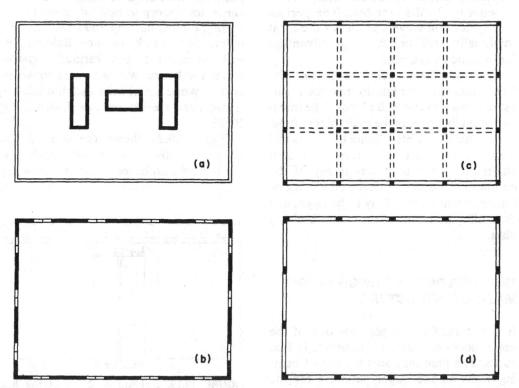

FIGURE 10.17. Building 3: alternative schemes for the lateral bracing system: (*a*) core system; (*b*) perimeter shear walls; (*c*) fully developed rigid frame; (*d*) perimeter rigid frame.

of rigid bents. For this building there would thus be four bents for bracing in one direction and five for bracing in the other direction. This requires that the beam-to-column connections be moment resistive.

6. *Peripheral Rigid-Frame System* (Fig. 10.17*d*). This consists of using only the columns and beams in the exterior walls, resulting in only two bracing bents in each direction.

In the right circumstances any of these systems may be acceptable. Each has advantages and disadvantages from both structural design and architectural planning points of view. The core-braced schemes were popular in the past, especially for buildings in which wind was the major concern. The core system allows for the greatest freedom in planning the exterior walls, which are obviously of major concern to the architect. The peripheral system, however, produces the most torsionally stiff building—an advantage for seismic resistance.

The rigid-frame schemes permit the free planning of the interior and the greatest openness in the wall planes. The integrity of the bents must be maintained, however, which restricts column locations and planning of stairs, elevators, and duct shafts so as not to interrupt any of the column-line beams. If designed for lateral forces, columns are likely to be large, and thus offer more intrusion in the building plan.

10.7 BUILDING 3: DESIGN OF THE MASONRY STRUCTURE

A structural framing plan for one of the upper floors of building 3 is shown in Fig. 10.18. The plan indicates the use of bearing walls as the major supports for the floor framing. The walls also constitute the lateral bracing system, with some

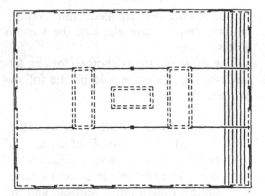

FIGURE 10.18. Building 3: framing plan for the masonry wall structure.

combination of a perimeter and core-braced system. There are many options for the floor framing, depending on fire-code requirements and the local competitive pricing of suppliers. For the office building, there are also many detailed concerns for incorporation of elements for wiring, piping, heating and cooling, ventilation, fire sprinklers, and lighting. We will assume that the various considerations can be met with a system consisting of a plywood deck, light nailable joists or trusses, and steel beams, as shown in Fig. 10.19.

Figure 10.20 shows the general construction of the exterior walls, indicating the use of reinforced concrete masonry.

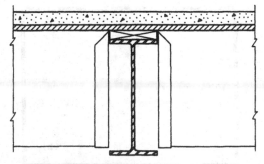

FIGURE 10.19. Building 3: detail of the floor structure with steel beams, plywood deck, wood joists, concrete topping.

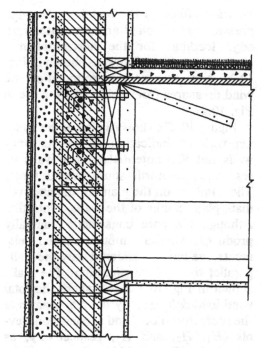

FIGURE 10.20. Building 3: detail of the exterior wall construction with exterior insulation system.

An exterior insulation system with applied finish is used on the outside surface and furring strips with gypsum drywall on the inside. The remainder of this discussion deals with the design of the structural masonry walls.

Design for gravity loads is relatively simple. There are two principal concerns: the general bearing in the walls, and concentrated loads from supported beams. Vertical gravity loads in the walls are greatest in the first story (ground level), so the masonry construction must facilitate this loading condition. For the reinforced concrete masonry there is a minimum construction that satisfies code requirements, typically requiring that at least every 4 ft of plan length of the wall have one concrete-filled, steel-reinforced void space. Because the voids are typically 8 in. on center, this means at least every sixth void will be reinforced. From that minimum, it is possible to upgrade the

wall strength by filling additional voids, and possibly by using more than the specified minimum reinforcement.

It is also possible to vary the wall thickness, which is more commonly done with *unreinforced* construction. However, for the three-story building, it is possible—and probably more feasible—to use a single-thickness wall for all stories, with use of the minimum construction at the top and progressive upgrading with added reinforcement in lower stories.

As the framing plan in Fig. 10.18 shows, the interior steel beams and the lintels in the exterior walls are supported at ends or corners of the walls. These will automatically be locations of reinforced voids in this construction, and can most likely to developed to provide the necessary concentrated strengths. However, it is also possible to develop enlarged masonry elements (as pilasters) at these locations if they are structurally required.

Design for Lateral Forces

The masonry walls must also develop resistance to lateral loadings, in combination with the horizontal roof and floor diaphragms. The final critical design of the walls must consider the combined effects of gravity and lateral loads. We first consider the effects of wind, using the criteria given previously and the requirements of the 1988 edition of the *UBC* (Ref. 1).

It is quite common, when designing for both wind and seismic forces, to have some parts of the structure designed for wind and others for seismic effects. In fact, what is necessary is to analyze for both effects and to design each element of the structure for the condition that produces the greater effect. Thus the shear walls may be designed for seismic effects, the exterior walls and window glazing for wind, and so on.

For wind it is necessary to establish the design wind pressure, defined by the code as

$$p = C_e C_q q_s I$$

where C_e is a combined factor including concerns for the height above grade, exposure conditions, and gusts. From *UBC* Table 23-G, assuming exposure B:

C_e = 0.7 from 0 to 20 ft above grade

= 0.8 from 20 to 40 ft

= 1.0 from 40 to 60 ft

and C_q is the pressure coefficient. Using the projected are method (method 2), we find from *UBC* Table 23-H the following.

For vertical projected area:

C_q = 1.3 up to 40 ft above grade

= 1.4 over 40 ft

For horizontal projected area (roof surface):

C_q = 0.7 upward

The symbol q_s is the wind stagnation pressure at the standard measuring height of 30 ft. From *UBC* Table 23-F the q_s value for a speed of 80 mph is 17 psf.

For the importance factor I (*UBC* Table 23-K) we use a value of 1.0.

Table 10.1 summarizes the foregoing data for the determination of the wind pressures at the various height zones for building 2. For the analysis of the horizon-tal wind effect on the building, the wind pressures are applied and translated into edge loadings for the horizontal dia-phragms (roof and floors) as shown in Fig. 10.21. Note that we have rounded off the wind pressures from Table 10.1 for use in Fig. 10.21.

Figure 10.22a shows a plan of the build-ing with an indication of the masonry walls that offer potential as shear walls for resistance to north–south lateral force. The numbers on the plan are the approxi-mate plan lengths of the walls. Note that although the core construction actually produces vertical tubular-shaped ele-ments, we have considered only the walls parallel to the load direction. The walls shown in Fig. 10.22a will share the total wind load delivered by the diaphragms at the roof, third-floor, and second-floor lev-els (H_1, H_2, and H_3, respectively, as shown in Fig. 10.21b). Assuming the building to be a total of 122-ft wide in the east–west direction, the forces at the three levels are

$$H_1 = 195 \times 122 = 23,790 \text{ lb [106 kN]}$$

$$H_2 = 234 \times 122 = 28,548 \text{ lb [127 kN]}$$

$$H_3 = 227 \times 122 = 27,694 \text{ lb [123 kN]}$$

and the total wind force at the base of the shear walls is the sum of these loads, or 80,032 lb [356 kN].

Although the distribution of shared load to masonry walls is usually done on the basis of a more sophisticated analysis

Table 10.1   DESIGN WIND PRESSURES FOR BUILDING 2

Height Above Average Level of Adjoining Ground (ft)	C_e	C_q	Pressure,[a] p (psf)
0–20	0.7	1.3	15.47
20–40	0.8	1.3	17.68
40–60	1.0	1.4	23.80

[a] Horizontally directed pressure on vertical projected area: $p = C_e \times C_q \times 17$ psf.

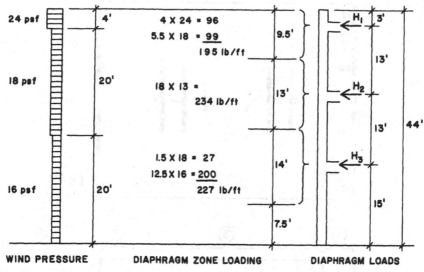

WIND PRESSURE   **DIAPHRAGM ZONE LOADING**   **DIAPHRAGM LOADS**

FIGURE 10.21. Building 3: development of wind loading to the roof and floor diaphragms.

for relative stiffness if we assume for the moment that the walls are stiff in proportion to their plan lengths (as is done with plywood walls), we may divide the maximum shear load at the base of the walls by the total of the wall plan lengths to obtain an approximate value for the maximum shear stress. Thus maximum shear is

$$v = \frac{80,032}{260}$$

$$= 308 \text{ lb/ft of wall length [4.49 kN/m]}$$

This is quite a low force for a reinforced masonry wall, which tells us that if wind alone is of concern we have considerable overkill in terms of total shear walls.

The shear stresses will not be equal in all walls because the lateral forces will not be distributed evenly between the walls. We visualize the distribution of the total lateral force to the walls by considering two extreme cases regarding the relative stiffness of the horizontal diaphragms (roof and floor desks).

First, if the horizontal diaphragms are considered to be infinitely stiff, then the distribution to individual walls will be in direct proportion to their individual relative stiffnesses. This method of distribution is illustrated in Sec. 5.8, using the distribution factors from the tables in Appendix C. Second, if the horizontal diaphragms are quite flexible, then the distribution to shear walls will be essentially on a load periphery basis.

Figure 10.22*b* shows the building plan with north–south shear and a breakdown of lateral load distribution on a load periphery basis. On this basis, the end shear walls carry a total of one quarter of the total load, and the combined core walls carry three quarters of the load. With this approach, the next step would be to consider the breakdown of load to each shear wall within the groups (ends and core).

Figure 10.22*c* indicates the relative stiffness of the individual walls with the three plan lengths of 10, 20, and 30 ft, considering the full wall height to the roof level. Using stiffness factors from Appendix C, the relative stiffnesses of the walls may be compared, resulting in the observation that the 10-ft walls have negligible

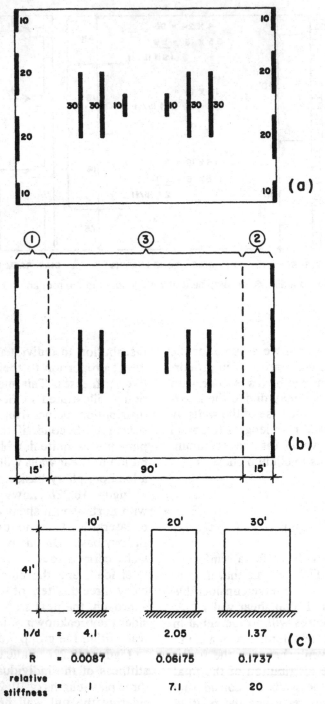

FIGURE 10.22. Building 3: considerations for distribution of lateral loads to the north–south shear walls.

effect in combination with the longer walls.

It is quite probable that the use of code-required minimum construction would result in a structure with adequate response to both gravity and wind loads for this building. Special attention would be required to ensure that the horizontal diaphragms are adequately anchored to the shear walls and that the foundations can develop the necessary resistances to sliding and overturn of the shear walls.

Investigation for seismic response for this building is likely to produce more critical concerns. The weight and relative stiffness of the masonry construction will result in major seismic loadings in seismic zones of high risk.

Appendix A

INVESTIGATION OF COMPRESSION ELEMENTS

The single major structural use of masonry is for the resistance of compression. Compression is developed in a number of ways in structures, including the compression component that accompanies the development of internal bending moment and the diagonal compression due to shear. In this section we consider elements whose primary purpose is the resistance of direct compression, although they may also be required to perform other structural tasks. The special considerations for composite elements of steel plus either concrete or masonry are treated in Appendix B.

A.1. TYPES OF COMPRESSION ELEMENTS

A number of types of primary compression elements are used in building structures. Major ones are the following.

Columns

Columns are usually linear vertical elements, used when supported loads are concentrated, or when a need for open space precludes the use of walls as supports. Relative slenderness may vary over a considerable range, from stout to slender, depending on the magnitude of the loads or the material and construction of the column. In various situations columns may also be called posts, piers, pedestals, or struts.

Piers

The term *pier* generally refers to relatively stout columns. The term is also used, however, to describe massive bridge supports, abutments, deep foundation elements cast in excavated shafts, and vertical masonry elements that are transitional in form between walls and columns. All of these elements usually resist major compression forces, but may also be required to develop other resistances, such as bending, lateral shear, or uplift.

Bearing Walls

When walls are used for supports, taking vertical compression, they are called bearing walls. If they are interior walls, this may be their singular structural function and they perform essentially like columns. Exterior walls, however, are usually also designed for major lateral bending due to wind or earth pressure or for action as shear walls, resisting shear effects in the wall plane.

A.2 SHORT COMPRESSION ELEMENTS

The general case for axial compression capacity as related to slenderness is indicated in Fig. A.1. The limiting conditions

are those of the very stout (not slender, usually meaning "short") element that fails essentially by crushing of the material and the very slender element that has its failure precipitated by buckling. Piers and abutments are typically of the short type, but columns or truss members may also fall in this range on occasion.

When subjected primarily to axial compression force, the capacity of the short compression member is directly proportional to the mass of the material and its strength in resisting crushing stresses, Fig. A.1, zone 1. The crushing limit may be established by column-type action, involving a generally uniformly distributed compression on the entire member cross section. However, as sometimes occurs with a pier or abutment, the transfer of the compressive load to the member may involve highly concentrated bearing stresses on some fraction of the whole member cross section. In the latter case, the load limit may be established by the compressive bearing stresses, not by the column action of the member.

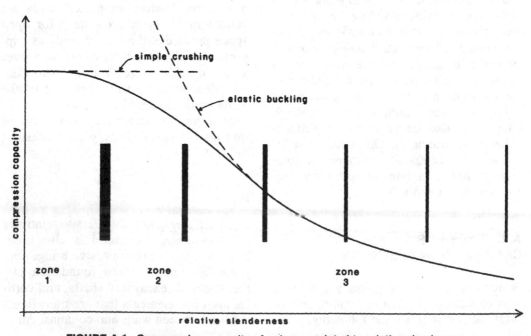

FIGURE A.1. Compression capacity of columns related to relative slenderness.

A.3 SLENDER COMPRESSION ELEMENTS

Very slender elements that sustain compression tend to buckle, Fig. A.1, zone 3. Buckling is a sudden lateral deflection, at right angles to the direction of the compression. If the member is held in position, the buckling may serve to relieve the member of the compressive effort and the member may spring or snap back into alignment when the compressive force is removed. If the force is not removed, the member will usually quickly fail—essentially by excessive bending action. It is essentially the inability of the member to resist significant bending that is the reason for its susceptibility to buckling.

The classic means for describing elastic buckling is the Euler curve, having the form

$$P = \frac{\pi^2 EI}{L^2}$$

This second-degree curve is one of the boundary conditions shown in Fig. A.1. From the form of the equation for the Euler curve, it may be noted that the pure elastic buckling response is

1. Proportional to the stiffness of the material of the member—indicated by E, the elastic modulus of elasticity of the material for direct stress.
2. Proportional to the bending stiffness of the member as indicated by moment of inertia I of its cross-sectional area.
3. Inversely proportional to the member length, or actually to the second power of the length. The length in this case is an indication of potential slenderness.

The two basic limiting response mechanisms—crushing and buckling—are entirely different in nature, relating to different properties of the material and of the form of the member. The crushing limit is indicated by the straight line on the graph in Fig. A.1; it is a constant value over the range of length it affects. The buckling load varies over the range of the member length, going to infinity for short length (which it does not actually affect) and to zero for extremely great length.

Buckling may be affected by constraints, such as lateral bracing that effectively prevents the sideways deflection, or end connections that restrain the rotation associated with the assumption of a single-curvature mode of deflection. Figure A.2a shows the case for the member that is the general basis for the response as indicated by the Euler formula. This form of response may be altered by lateral constraints, as shown in Fig. A.2b, that result in a multimode response. In the example shown in Fig. A.2b, the deflected form is such that the length over which buckling occurs is reduced to one-half of the column height.

In comparison to the member in Fig. A.2a, the member in Fig. A.2c has its ends restrained against rotation, having what is described as fixed ends. In this case the deflected (buckled) shape is a doubly inflected curve. It can be shown that the inflection points of this curve will occur at the quarter points of the height; thus the length over which the simple free buckling occurs is one half the full height of the member. Conditions such as all three cases in Fig. A.2 occur in building structures, and modifications are used to

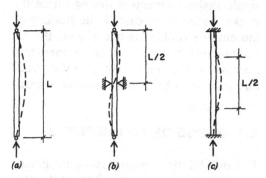

FIGURE A.2. Effect of restraining conditions on buckling of columns.

qualify the member response on the basis of evaluation of whatever constraints are judged to occur. One means of qualifying the response is by using an *effective buckling length*. In the cases shown in Fig. A.2*b* and *c* this modified length would be one-half of the actual length.

Member straightness in the unloaded condition can be a critical factor for the very slender member. If the member is not reasonably straight, it may be in what is virtually a prebuckled condition and buckling may occur progressively with a minor load. A deflected position can also occur due to bending induced by other loadings (Secs. A.5 and A.6).

A final consideration for the very slender member is the relative dynamic character of the load. Buckling is essentially a dynamic type of response, occurring as a very sudden collapse. This is not a very desirable mode of failure for building structures. Not that any failure is desirable; yet there are those that have a nature that is better than others. Brittle fracture and buckling are less preferred than slow ductile yielding or multistaged responses. If load buildup is slow, a buckling tendency may actually be detected in time for corrective measures to be taken (provide bracing or remove load). If the load buildup is quick, however, buckling will most likely occur suddenly and without warning. Truly dynamic loads are often not of a long-time occurrence, such as a surge of wind pressure (called a gust) or a single major movement during an earthquake. In the latter case, if the stability of the entire structure is not at risk, the instantaneous buckling of some members may not result in collapse, and the structure may snap back into a safe condition.

A.4 RANGE OF SLENDERNESS

The two limiting responses—pure crushing and pure elastic buckling—are actually only true at the extreme ends of the range of member length or slenderness. Between these limits there is a transitional range in which the actual response is some combination of the two forms of response. Except for piers and abutments, most primary compression members in building structures fall in this range. The three ranges, numbered 1, 2, and 3, are indicated on Fig. A.1. The form of the response on the graph shows the behavior in the intermediate range (zone 2) to be simply a geometric transition from the straight horizontal line representing crushing to the second-degree curve of the Euler formula. The points of actual transition are arbitrary in the illustration, but have been repeatedly demonstrated in laboratory tests.

For the design of steel and wood columns, codes provide criteria for the evaluation of compression capacity over the full range of slenderness. For masonry and concrete columns some adjustment of capacity is also provided for; however, columns in these materials are actually mostly quite stout.

On Fig. A.1 the proportions of columns corresponding to the range of slenderness are indicated. These are actually derived from present code criteria for solid wood columns of ordinary structural-grade lumber. Some minor variation will occur with other materials and with other cross-sectional forms, but the effect on the column profiles will be small. Consideration of these profiles will confirm our earlier allegation that most building structural columns tend to fall in the intermediate range.

A.5 INTERACTION: COMPRESSION PLUS BENDING

There are a number of situations in which structural members are subjected to the combined effects that result in develop-

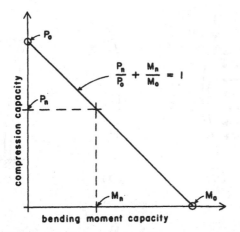

FIGURE A.3. Column interaction: compression plus bending.

ment of axial compression and internal bending. Stresses developed by these two actions are both of the direct stress type and can be directly combined for the consideration of net stress conditions. The stress investigation is considered in Sec. A.9. However, the actions of a column and a bending member are essentially different in character and it is therefore customary to consider this combined activity by what is described as interaction.

The classic form of interaction is represented by the graph in Fig. A.3. Referring to the notation on the graph:

The maximum axial load capacity of the member (with no bending) is P_o.

The maximum bending moment capacity of the member (without compression) is M_o.

At some compression load below P_o (indicated as P_n) the member is assumed to have some tolerance for a bending moment (indicated as M_n) in combination with the axial load.

Combinations of P_n and M_n are assumed to fall on a line connecting P_o and M_o. The equation of this line has the form

$$\frac{P_n}{P_o} + \frac{M_n}{M_o} = 1$$

A graph similar to that in Fig. A.3 can be constructed using stresses rather than loads and moments. This is the procedure used with wood and steel members; the graph takes a simple form expressed as

$$\frac{f_a}{F_a} + \frac{f_b}{F_b} \leq 1$$

where

f_a = actual stress due to the axial load
F_a = allowable column-action stress
f_b = actual stress due to bending
F_b = allowable beam-action stress in flexure

For various reasons, real structural members do not adhere to the classic straight-line form of response as shown for interaction in Fig. A.3. Figure A.4 shows a form of response characteristic of reinforced concrete columns. While there is some approximation of the pure interaction response in the midrange of combined effects, major deviation occurs at both ends of the range. At the low moment end, where almost pure compression occurs, the column is capable of developing only some percentage of the full, theoretical

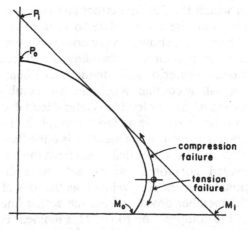

FIGURE A.4. Interaction response of a reinforced concrete column; general form.

capacity. There are many contributing reasons for this, including the general nature of the composite (concrete and steel) material and typical inaccuracies in construction. At the high end of the moment range, where little compression occurs, failure is controlled by various procedures and criteria that tend to result in a failure that is predicated by tension yielding of the reinforcing, rather than by compression failure of the concrete. As compression is added at the low end of the axial load range, it actually tends to increase the moment capacity, up to the point where failure eventually becomes one essentially of compression.

Steel and wood members also have various deviations from the simple interaction form of response. A major effect is the so-called *P*-delta, discussed in the next section. Other problems include inelastic behavior, effects of lateral stability, and general effects of the geometry of member cross sections.

A.6   THE *P*-DELTA EFFECT

Bending moments can be developed in structural members in a number of ways. When the member is subjected to an axial compression force, there are various ways in which the bending effect and the compression effect can relate to each other. Figure A.5*a* shows a very common situation that occurs in building structures when an exterior wall functions as a bearing wall or contains a column. The combination of gravity load plus lateral load due to wind or seismic action can result in the loading shown. If the member is quite flexible, and the actual deflection from the unloaded position is significant, an additional moment is developed as the axis of the member deviates from the action line of the compression force. This moment is the simple product of the load and the member deflection at any point; that is *P*

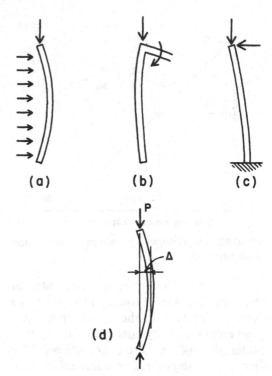

FIGURE A.5. Sources of bending in columns: (a) lateral load from wind, seismic, or soil pressure effects; (b) transferred moment in rigid frames; (c) lateral load on a cantilevered column; (d) the *P*-delta effect of an axial compression load on a deflected column.

times Δ, as shown in Fig. A.5*d*. It is thus referred to as the *P*-delta effect.

There are various other situations that can result in this effect. Figure A.5*b* shows an end column in a rigid frame structure, where moment is induced at the top of the column by the moment-resistive connection to the beam. Although slightly different in its geometric profile, the basic column response is similar to that in Fig. A.4*a* unless, of course, the frame is also subjected to a sideways deflection due to a lateral load or some lack of symmetry in the frame. Figure A.5*c* shows the effect of a combination of gravity and lateral loads on a vertically cantilevered structure that supports a sign or a tank at its top. The two effects shown in Fig. A.5*b* and *c* will be combined when a rigid frame is also subjected to the combined vertical and lateral loads.

In all of these, and various other, situations, the *P*-delta effect may or may not be critical. The major factor that determines its seriousness is the relative flexibility of the structure in general, but particularly the stiffness of the member that directly sustains the effect. In a worst-case scenario, the *P*-delta effect may be an accelerating one, in which the added moment due to the *P*-delta effect causes additional deflection, which in turn results in additional *P*-delta effect, which then causes more deflection, and so on.

The *P*-delta effect is seldom a solitary condition, except for the case of the member with a significant lack of initial straightness. In most cases, the effect will be combined with other conditions. For the slender element, the *P*-delta effect may work to precipitate a buckling failure. In other cases the moment due to the *P*-delta effect may simply combine with other moments, and the resulting action becomes one of interaction or combined stress as a critical condition.

There are actually not very many real situations in which the *P*-delta effect is critical. Although it occurs and adds to other conditions in many situations, it is most often only a minor effect. The potentially critical situations are usually those involving super-slender structural members, particularly very tall or very skinny columns. In the latter case, the *P*-delta effect should be carefully considered, regardless of the magnitude of other effects.

A.7 COMPRESSION IN CONFINED MATERIALS

Solid materials have capability to resist linear compression effects. Fluids can resist compression only if they are in a confined situation, such as air in an auto tire or oil in a hydraulic jack. Compression of a confined material results in a three-dimensional compressive stress condition, visualized as a triaxial condition, as shown in Fig. A.6.

A major occurrence of the triaxial stress condition is that which exists in loose soil materials. Supporting materials for foundations are typically buried below some amount of soil overburden. The upper mass of soil, plus the general effect of the surrounding soil, creates a considerable confinement. This confinement permits some otherwise compression-weak materials to accept some amount of compression force. Loose sands and soft, wet clays must have this confinement, although they are still not very desirable bearing materials, even with the confinement. The confinement is a necessary continuing condition for their stability, and removal or reduction due to effects such as excavating near the soil mass, or a drop in the groundwater level, may cause problems.

While confinement is mandatory for fluid or loose, granular materials, it can also enhance the compression resistance of solids. An example of this is the concrete in the center of a reinforced concrete spiral column. The spiral column utilizes a continuous helix (in the general shape of a coiled spring) which is near the column perimeter. At loads near the ultimate capacity of the column, the spiral will de-

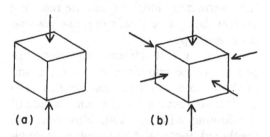

FIGURE A.6. Conditions of developed compressive stress: (*a*) simple direct compression; (*b*) compression in a confined material–triaxial stress.

velop tension and act to confine the concrete in the center mass of the column. This confined concrete will develop a level of compression higher than that of which it is capable in a simple linear stress situation.

A.8 COMBINED STRESS: COMPRESSION PLUS BENDING

Combined actions of compression and bending produce various effects on structures. The general interaction of the two separate phenomena and possible concern for P-delta effects are discussed in earlier sections of this chapter. We will now consider the condition of the actual stress combinations that occur when axial compression is added to bending moment at some cross section. One common example of this is the development of stress on the bottom of a bearing footing; the "section" in this case being the bearing contact face of the footing bottom.

Figure A.7 shows a situation in which a simple rectangular footing is subjected to a combination of forces that require the resistance of vertical force, horizontal sliding, and overturning moment. The development of resistance to horizontal movement is produced by some combination of friction on the bottom of the footing and horizontal earth pressure on the face of the footing. Our concern here is for the investigation of the vertical force and the overturning moment and the resulting combination of vertical soil pressures that they develop.

Figure A.7 illustrates our usual approach to the combined direct force and moment on a cross section. In this case the "cross section" is the contact face of the footing with the soil. However, the combined force and moment may originate, we make a transformation into an equivalent eccentric force that produces the same effects on the cross section. The direction and magnitude of this mythical equivalent e are related to properties of the cross section in order to qualify the nature of the stress combination. The value of e is established simply by dividing the moment by the force normal to the cross section, as shown in the figure. The net, or combined, stress distribution on the section is visualized as the sum of the separate stresses due to the normal force and the moment. For the stresses on the two extreme edges of the footing the general formula for the combined stress is

$$p = \frac{N}{A} \pm \frac{Nec}{I}$$

We observe three cases for the stress combination obtained from this formula, as shown in the figure. The first case occurs when e is small, resulting in very little bending stress. The section is thus subjected to all compressive stress, varying from a maximum value on one edge to a minimum on the opposite edge.

The second case occurs when the two stress components are equal, so that the minimum stress becomes zero. This is the boundary condition between the first and third cases, since any increase in the eccentricity will tend to produce some tension stress on the section. This is a significant limit for the footing since tension stress is not possible for the soil-to-footing contact face. Thus case 3 is possible only in a beam or column where tension stress can be developed. The value of e that corresponds to case 2 can be derived by equating the two components of the stress formula as follows:

$$\frac{N}{A} = \frac{Nec}{I}, \qquad e = \frac{I}{Ac}$$

This value for e establishes what is called the *kern limit* of the section. The kern is a zone around the centroid of the section within which an eccentric force will not

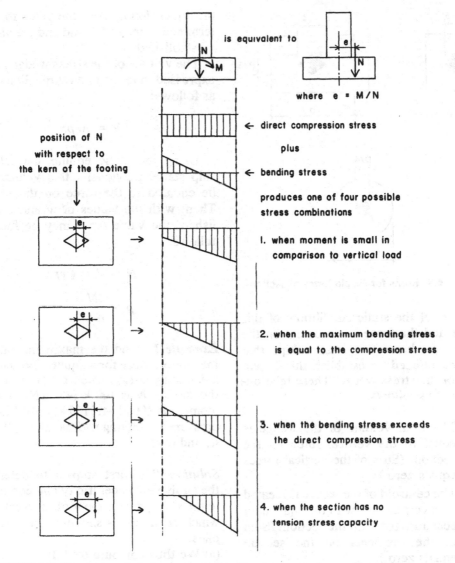

is equivalent to

where e = M / N

position of N
with respect to
the kern of the footing

← direct compression stress

plus

← bending stress

produces one of four possible
stress combinations

I. when moment is small in
comparison to vertical load

2. when the maximum bending stress
is equal to the compression stress

3. when the bending stress exceeds
the direct compression stress

4. when the section has no
tension stress capacity

FIGURE A.7: Development of stress due to combined compression and bending at a section.

cause tension on the section. The form of this zone may be established for any shape of cross section by application of the formula derived for the kern limit. The forms of the kern zones for three common shapes of section are shown in Fig. A.8.

When tension stress is not possible, eccentricities beyond the kern limit will produce a so-called *cracked section*, which is shown as case 4 in Fig. A.7. In this situa-

tion some portion of the section becomes unstressed, or cracked, and the compressive stress on the remainder of the section must develop the entire resistance to the force and moment.

Figure A.9 shows a technique for the analysis of the cracked section, called the *pressure wedge method*. The pressure wedge represents the total compressive force developed by the soil pressure.

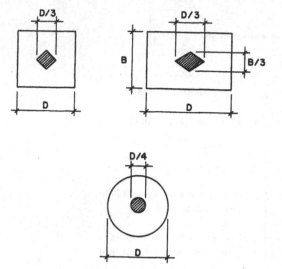

FIGURE A.8. Kerns for simple forms of sections.

Analysis of the static equilibrium of this wedge and the force and moment on the section produces two relationships that may be utilized to establish the dimensions of the stress wedge. These relationships are as follows:

1. The total volume of the wedge is equal to the vertical force on the section. (Sum of the vertical forces equals zero.)
2. The centroid of the wedge is located on a vertical line with the equivalent eccentric force on the section. (Sum of the moments on the section equals zero.)

Referring to Fig. A.9, the three dimensions of the stress wedge are w, the width of the footing, p, the maximum soil pressure, and x, the limit of the uncracked portion of the section. With w known, the solution of the wedge analysis consists of determining values for p and x. For the rectangular footing, the simple triangular stress wedge will have its centroid at the third point of the triangle. As shown in the figure, this means that x will be three times

the dimension a. With the value for e determined, a may be found and the value of x established.

The volume of the stress wedge may be expressed in terms of its three dimensions as follows:

$$V = \tfrac{1}{2}wpx$$

Using the static equilibrium relationship stated previously, this volume may be equated to the force on the section. Then, with the values of w and x established, the value for p may be found as follows:

$$N = V = \tfrac{1}{2}wpx$$

$$p = \frac{2N}{wx}$$

Example 1. Find the maximum value of the soil pressure for a square footing. The axial compression force at the bottom of the footing N is 100 k [450 kN], and the moment is 100 kip-ft [135 kN-m]. Find the pressure for footing widths of (a) 8 ft, (b) 6 ft, and (c) 5 ft.

Solution: The first step is to determine the equivalent eccentricity and compare it to the kern limit for the footing to establish which of the cases shown in Fig. A.7 applies.

(a) We thus compute for 8 ft:

$$e = \frac{M}{N} = \frac{100}{100} = 1 \text{ ft } [0.3 \text{ m}]$$

Kern for the 8-ft wide footing $= \dfrac{8}{6}$

$$= 1.33 \text{ ft } [0.41 \text{ m}]$$

and it is established that case 1 applies.

We next determine the soil pressure, using the formula for the combined stress as previously derived.

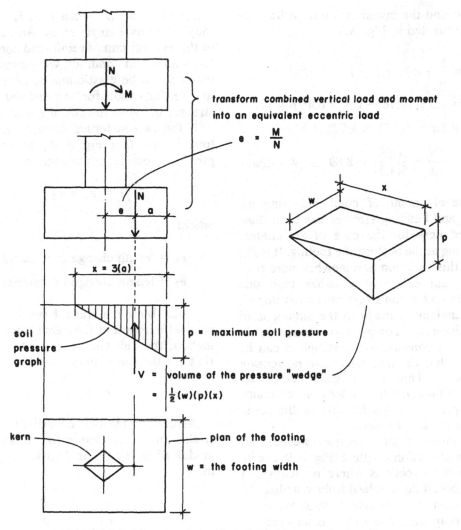

transform combined vertical load and moment
into an equivalent eccentric load

$$e = \frac{M}{N}$$

p = maximum soil pressure

soil pressure graph

V = volume of the pressure "wedge"

$$= \frac{1}{2}(w)(p)(x)$$

kern

plan of the footing

w = the footing width

FIGURE A.9. The pressure-wedge method for a cracked section.

$$p = \frac{N}{A} + \frac{Mc}{I} = \frac{100}{64} + \frac{100 \times 4}{341.3}$$

$$= 1.56 + 1.17 = 2.73 \text{ ksf}$$

$$[75.6 + 56.1 = 131.7 \text{ } kPa]$$

in which

$$A = (8)^2 = 64 \text{ ft}^2 \text{ } [5.95 \text{ m}^2]$$

$$I = \frac{bd^3}{12} = \frac{(8)^4}{12} = 341.3 \text{ ft}^4 \text{ } [2.95 \text{ m}^4]$$

(b) It may be observed that the kern limit is 6/6 = 1, which is equal to the eccentricity. Thus the situation is that shown as case 2 in Fig. A.7, and the pressure is such that $N/A = Mc/I$. Thus

$$p = 2\left(\frac{N}{A}\right) = 2\left(\frac{100}{6(6)}\right)$$

$$= 5.56 \text{ ksf } [266 \text{ kPa}]$$

(c) The eccentricity exceeds the kern

limit, and the investigation must be done as illustrated in Fig. A.9.

$$a = \frac{5}{2} - e = 2.5 - 1$$

$$= 1.5 \text{ ft } [0.76 - 0.3 = 0.46 \text{ m}]$$

$$x = 3a = 3(1.5) = 4.5 \text{ ft } [1.38 \text{ m}]$$

$$p = \frac{2N}{wx} = \frac{2(100)}{5(4.5)} = 8.89 \text{ ksf } [429 \text{ kPa}]$$

Development of combined compression and bending stresses has been illustrated here for the case of the contact bearing at the bottom of a footing. It is felt that this situation is somewhat more realistic and easier to visualize than one where no tension stress can be developed. An analogy is made to the situations of unreinforced concrete or masonry, in which a conservative assumption can be made that the materials have no tension capacity. This is not really completely true, although ultimate tension resistance is typically only a fraction of the resistance to compression.

Although the pressure-wedge, or cracked-section, method (Fig. A.9) can be applied to sections where no tension is possible, it does indeed truly visualize the development of an actual crack. Since it is generally desirable to eliminate cracking in masonry and concrete structures, it is advisable to consider stress combination No. 2 in Fig. A.7 (zero tension at the edge of the section) as the real limit for good design. However, for the extreme load conditions of major wind or earthquake effects, it may be reasonable to use the cracked section for determination of an ultimate resistance.

A.9 COMPOSITE ELEMENTS

A special stress condition occurs when two or more materials are assembled in a single unit so that when load is applied they strain as a single mass. An example of this is a column of reinforced concrete. In an idealized condition we assume both materials to be elastic and make the following derivation for the distribution of stresses between the two materials.

If the two materials deform the same total amount (e in Fig. A.10), we may express the total length change as

$$e = e_1 = e_2$$

where

e_1 = length change of material 1

e_2 = length change of material 2

Since both materials have the same original length and the same total deformation, the unit strains in the two materials are the same; thus

$$\varepsilon_1 = \varepsilon_2$$

Assuming elastic conditions, these strains may be related to the stresses and moduli of elasticity for the two materials; thus

$$\varepsilon_1 = \frac{f_1}{E_1} = \varepsilon_2 = \frac{f_2}{E_2}$$

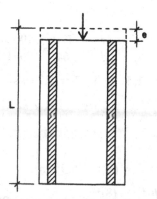

FIGURE A.10. Shared compression in a composite element.

The relationship of the stresses in the two materials may thus be stated as

$$\frac{f_1}{f_2} = \frac{E_1}{E_2} \quad \text{or} \quad f_1 = f_2 \frac{E_1}{E_2}$$

Expressed in various ways, the relationship is simply that the stresses in the materials are proportional to their moduli of elasticity.

Example 1. A reinforced concrete column consists of a 12-in square concrete section with four 0.75-in.-diameter round steel rods. The column sustains a compression load of 100 kips. Find the stresses in the concrete and steel. (Assume an E of 4000 ksi for concrete and 29,000 ksi for steel.)

Solution: Consider the load to be resisted by two internal forces, P_s and P_c—that is, the load resisted by the steel and the load resisted by the concrete. Then

$$\text{Total } P = P_s + P_c = f_s A_s + f_c A_c$$

Using the previously derived relationship for the two stresses yields

$$f_s = \frac{E_s}{E_c} f_c = \frac{29,000}{4000} f_c = 7.25 f_c$$

Substituting this in the expression for P yields

$$\begin{aligned} P = 100 &= f_s A_s + f_c A_c \\ &= 7.25 f_c A_s + f_c A_c \\ &= f_c (7.25 A_s + A_c) \end{aligned}$$

For one steel bar

$$A = \pi R^2 = 3.14(0.375)^2 = 0.44 \text{ in.}^2$$

and for all four bars

$$A_s = 4(0.44) = 1.76 \text{ in.}^2$$

Then the concrete area is

$$A_c = (12)^2 - 1.76 = 142.24 \text{ in.}^2$$

Substituting values yields

$$100 = f_c\{7.25(1.76) + 142.24\} = f_c(155.0)$$

Thus

$$f_c = \frac{100}{155.0} = 0.645 \text{ ksi}$$

and

$$f_s = 7.25 f_c = 7.25(0.629) = 4.68 \text{ ksi}$$

INVESTIGATION AND DESIGN OF REINFORCED CONCRETE: WORKING-STRESS METHOD

The following is a brief presentation of the formulas and procedures used in the working-stress method. It is recommended that this method be used only for approximate designs for preliminary studies or for designs with low strength concrete ($f_c' = 3000$ psi or less) and low amounts of reinforcement (percent of steel of 1.0 or less).

B.1 FLEXURE—RECTANGULAR SECTION WITH TENSION REINFORCEMENT ONLY

The symbols in Fig. B.1 are defined as follows:

b = width of the concrete compression zone

d = effective depth of the section for stress analysis; from the centroid of the steel to the edge of the compression zone

A_s = cross-sectional area of the reinforcing

p = percentage of reinforcing, defined as

$$p = \frac{A_s}{bd}$$

n = elastic ratio

$$= \frac{E \text{ of the steel reinforcing}}{E \text{ of the concrete}}$$

kd = height of the compression stress zone; used to locate the neutral axis of the stressed section; expressed as a percentage (k) of d

jd = internal moment arm between the net tension force and the net compression force; expressed as a percentage (j) of d

f_c = maximum compressive stress in the concrete

f_s = tensile stress in the reinforcing.

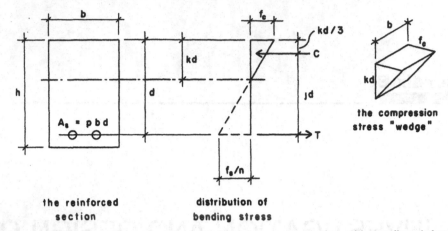

FIGURE B.1. Development of bending in a reinforced concrete section with tensile reinforcement: working-stress method.

The compression force C may be expressed as the volume of the compression stress "wedge," as shown in the figure.

$$C = \tfrac{1}{2}(kd)(b)(f_c) = \tfrac{1}{2}kf_cbd$$

Using the compression force, the moment resistance of the section may be expressed as

$$M = Cjd = (\tfrac{1}{2}kf_cbd)(jd) = \tfrac{1}{2}kjf_cbd^2 \quad (1)$$

This may be used to derive an expression for the concrete stress:

$$f_c = \frac{2M}{kjbd^2} \quad (2)$$

The resisting moment may also be expressed in terms of the steel and the steel stress as

$$M = Tjd = (A_s)(f_s)(jd)$$

This may be used for determination of the steel stress or for finding the required area of steel:

$$f_s = \frac{M}{A_sjd} \quad (3)$$

$$A_s = \frac{M}{f_sjd} \quad (4)$$

A useful reference is the so-called balanced section, which occurs when the exact amount of reinforcing used results in the simultaneous limiting stresses in the concrete and steel. The properties that establish this relationship may be expressed as follows:

$$\text{Balanced } k = \frac{1}{1 + f_s/nf_c} \quad (5)$$

$$j = 1 - \frac{k}{3} \quad (6)$$

$$p = \frac{f_ck}{2f_s} \quad (7)$$

$$M = Rbd^2 \quad (8)$$

in which

$$R = \tfrac{1}{2}kjf_c \quad (9)$$

[derived from formula (1).] If the limiting compression stress in the concrete ($f_c = 0.45f_c'$) and the limiting stress in the steel are entered in formula (5), the balanced section value for k may be found. Then the corresponding values for j, p, and R may be found. The balanced p may be used to determine the maximum amount of tensile reinforcing that may be used in a section without the addition of compressive reinforcing. If less tensile reinforcing is used, the moment will be limited by the steel stress, the maximum stress in the concrete will be below the limit of $0.45f'_c$, the value of k will be slightly lower than the balanced value, and the value of j slightly higher than the balanced value. These relationships are useful in design for the determination of approximate requirements for cross sections.

Table B.1 gives the balanced section properties for various combinations of concrete strength and limiting steel stress. The values of n, k, j, and p are all without units. However, R must be expressed in particular units; the units used in the table are kip-inches (k-in.) and kilonewton-meters (kN-m).

When the area of steel used is less than the balanced p, the true value of k may be determined by the formula

$$k = \sqrt{2np - (np)^2} - np \qquad (10)$$

Figure B.2 may be used to find approximate k values for various combinations of p and n.

B.2 REINFORCED CONCRETE COLUMNS

The practicing structural designer customarily uses tables or a computer-aided procedure to determine the dimensions and reinforcing for concrete columns. The complexity of analytical formulas and the large number of variables make it impractical to perform design for a large number of columns solely by hand computation. The provisions relating to the design of columns in the 1988 ACI Code are quite different from those of the working-stress design method in the 1963 Code. The current code does not permit design of columns by the working-stress method, but it rather requires that the service load capacity of columns be determined as 40% of that computed by strength design procedures.

Due to the nature of most concrete structures, current design practices generally do not consider the possibility of a concrete column with axial compression alone. That is to say, the existence of some bending moment is always considered together with the axial force. Figure B.3a illustrates the nature of the so-called *interaction response* for a concrete column with a range of combinations of axial load plus bending moment. In gen-

Table B.1. BALANCED SECTION PROPERTIES FOR RECTANGULAR CONCRETE SECTIONS WITH TENSION REINFORCING ONLY

| f_s | | f'_c | | | | | | R | |
ksi	MPa	ksi	MPa	n	k	j	p	k-in.	kN-m
16	110	2.0	13.79	11.3	0.389	0.870	0.0109	0.152	1045
		2.5	17.24	10.1	0.415	0.862	0.0146	0.201	1382
		3.0	20.68	9.2	0.437	0.854	0.0184	0.252	1733
		4.0	27.58	8.0	0.474	0.842	0.0266	0.359	2468
20	138	2.0	13.79	11.3	0.337	0.888	0.0076	0.135	928
		2.5	17.24	10.1	0.362	0.879	0.0102	0.179	1231
		3.0	20.68	9.2	0.383	0.872	0.0129	0.226	1554
		4.0	27.58	8.0	0.419	0.860	0.0188	0.324	2228
24	165	2.0	13.79	11.3	0.298	0.901	0.0056	0.121	832
		2.5	17.24	10.1	0.321	0.893	0.0075	0.161	1107
		3.0	20.68	9.2	0.341	0.886	0.0096	0.204	1403
		4.0	27.58	8.0	0.375	0.875	0.0141	0.295	2028

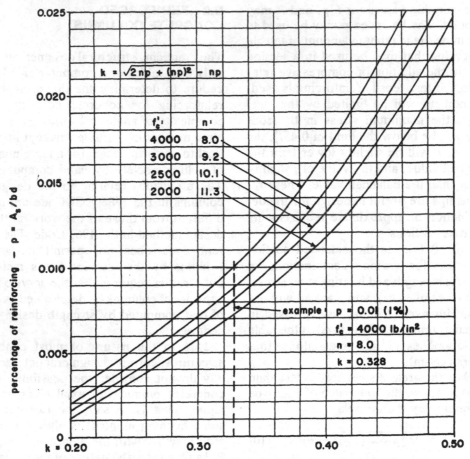

FIGURE B.2. *k* factors for rectangular concrete beams, as a function of *p* and *n*.

eral, the three basic ranges of this behavior are as follows:

1. *Large Axial Force, Minor Moment.* For this case, the moment has little effect, and the resistance to pure axial force is only negligibly reduced.

2. *Significant Values for Both Axial Force and Moment.* For this case, the analysis for design must include the full combined force effects, that is, the interaction of the axial force and the bending moment.

3. *Large Bending Moment, Minor Axial Force.* For this case, the column behaves essentially as a doubly rein-

forced (tension and compression reinforced) member, with its capacity for moment resistance affected only slightly by the axial force.

In Fig. B.3*a* the solid line on the graph represents the true response of the column—a form of behavior verified by many load tests on laboratory specimens. The dashed line figure on the graph represents the generalization of the three types of response just described.

The terminal points of the interaction response—pure axial compression or pure bending moment (P_o and M_o on the graph)—may be reasonably easily determined. The interaction responses between these two limits require complex analy-

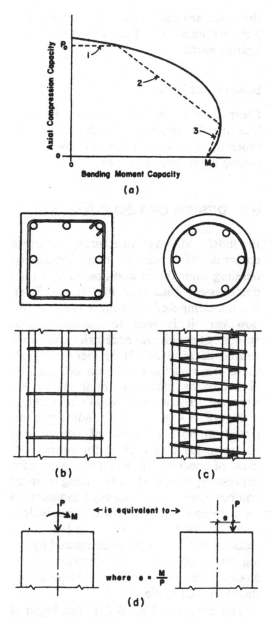

FIGURE B.3. Reinforced concrete columns: (a) form of the response to combined compression and bending; (b) typical tied column; (c) typical spiral column; (d) visualization of combined compression and bending as an equivalent eccentric force.

ses, which are beyond the scope of this book.

Reinforced concrete columns for buildings generally fall into one of the following categories:

1. Square tied columns
2. Round spiral columns
3. Oblong tied columns
4. Columns of other geometries (hexagonal, L-shaped, T-shaped, etc.) with either ties or spirals

In tied columns the longitudinal reinforcing is held in place by loop ties made of small-diameter reinforcing bars, commonly No. 3 or No. 4. Such a column is represented by the square section shown in Fig. B.3b. This type of reinforcing can quite readily accommodate other geometries as well as the square. The design of such a column is discussed in Sec. B.3.

In spiral columns the longitudinal reinforcing is placed in a circle, with the whole group of bars enclosed by a continuous cylindrical spiral made from steel rod or large-diameter steel wire. Although this reinforcing system obviously works best with a round column section, it can be used also with other geometries. A round column of this type is shown in Fig. B.3c.

Experience has shown the spiral column to be slightly stronger than an equivalent tied column with the same amount of concrete and reinforcing. For this reason, code provisions allow slightly more load on spiral columns. Spiral reinforcing tends to be expensive, however, and the round bar pattern does not always mesh well with other construction details in buildings. Thus tied columns are often favored where restrictions on the outer dimensions of the sections are not severe.

Code provisions and practical construction considerations place the following restrictions on column dimensions and choice of reinforcing.

Column Size

The current Code does not contain limits for column dimensions. For practical reasons, the following limits are recommended. Rectangular tied columns should be limited to a minimum area of 100 in.[2] and a side dimension of 10 in. if square and 8 in. if oblong. Spiral columns should be limited to a minimum size of 12 in. if either round or square.

Reinforcing

Minimum bar size is No. 5. The minimum number of bars is four for tied columns, five for spiral columns. The minimum amount of area of steel is 1% of the gross column area. A maximum area of steel of 8% of the gross area is permitted, but bar spacing limitations make this difficult to achieve; 4% is a more practical limit. Section 10.8.4 of the 1983 ACI Code stipulates that for a compression member with a larger cross section than required by considerations of loading, a reduced effective area not less than one-half the total area may be used to determine minimum reinforcement and design strength.

Ties

Ties shall be at least No. 3 for bars No. 10 and smaller. Number 4 ties should be used for bars that are No. 11 and larger. Vertical spacing of ties shall be not more than 16 times the bar diameter, 48 times the tie diameter, or the least dimension of the column. Ties shall be arranged so that every corner and alternate longitudinal bar is held by the corner of a tie with an included angle of not greater than 135°, and no bar shall be farther than 6 in. clear from such a supported bar. Complete circular ties may be used for bars placed in a circular pattern.

Concrete Cover

A minimum of 1.5 in. is needed when the column surface is not exposed to weather or in contact with the ground; 2 in. should be used for formed surfaces exposed to the weather or in contact with the ground; 3 in. are necessary if the concrete is cast against earth.

Spacing of Bars

Clear distance between bars shall not be less than 1.5 times the bar diameter, 1.33 times the maximum specified size for the coarse aggregate, or 1.5 in.

B.3   DESIGN OF TIED COLUMNS

In most building structures, concrete columns will sustain some computed bending moment in addition to the axial compression load (see Fig. B.3a). Even when a computed moment is not present, however, it is well to consider some amount of accidental eccentricity or other source of moment. It is recommended, therefore, that the maximum safe load be limited to that given for a minimum eccentricity of 10% of the column dimension.

Figure B.4 gives safe loads for a selected number of sizes of square tied columns. Loads are given for various degrees of eccentricity, which is a means for expressing axial load and bending moment combinations. The computed moment on the column is translated into an equivalent eccentric loading, as shown in Fig. B.3d. Data for the curves were computed by using 40% of the load determined by strength design methods, as required by the 1988 ACI Code.

The curves in Fig. B.4 do not begin at zero eccentricity. In addition, the requirement that only 40% of the load determined by strength methods be used places a rather high safety factor on the working stress method. To say the least, the Code does not favor the working stress method in this case.

The following examples illustrate the use of Fig. B.4 for the design of tied columns.

Example 1. A column with f'_c = 4 ksi and steel with f_y = 60 ksi sustains an axial compression load of 400 k. Find the minimum practical column size if reinforcing is a maximum of 4% and the maximum size if reinforcing is a minimum of 1%.

Solution: using Fig. B.4*b*, we find from the sizes given:

Minimum column is 20 in.² with eight No. 9 (curve No. 14).

Maximum capacity is 410 kips, p_g = 2.0%.

Maximum size is 24 in.² with four No. 11 (curve no. 17)

Maximum capacity is 510 kips, p_g = 1.08%.

It should be apparent that it is possible to use an 18-in. or 19-in. column as the minimum size and to use a 22-in. or 23-in. column as the maximum size. Since these sizes are not given in the figure, we cannot verify them for certain without using strength design procedures.

Example 2. A square tied column with f'_c = 4 ksi and steel with f_y = 60 ksi sustains an axial load of 400 kips and a bending moment of 200 kip-ft. Determine the minimum size column and its reinforcing.

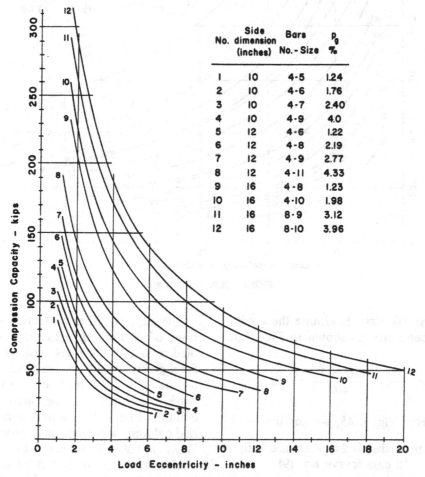

No.	Side dimension (inches)	Bars No.-Size	p_g %
1	10	4-5	1.24
2	10	4-6	1.76
3	10	4-7	2.40
4	10	4-9	4.0
5	12	4-6	1.22
6	12	4-8	2.19
7	12	4-9	2.77
8	12	4-11	4.33
9	16	4-8	1.23
10	16	4-10	1.98
11	16	8-9	3.12
12	16	8-10	3.96

FIGURE B.4. Safe service loads for square tied columns with specified concrete strength of 4 ksi and Grade 60 reinforcement.

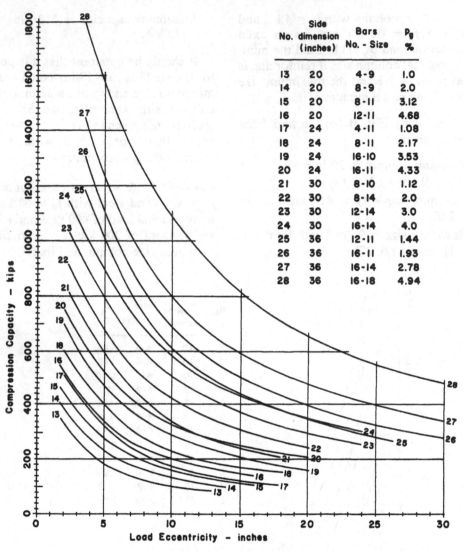

No.	Side dimension (inches)	Bars No. - Size	P_g %
13	20	4-9	1.0
14	20	8-9	2.0
15	20	8-11	3.12
16	20	12-11	4.68
17	24	4-11	1.08
18	24	8-11	2.17
19	24	16-10	3.53
20	24	16-11	4.33
21	30	8-10	1.12
22	30	8-14	2.0
23	30	12-14	3.0
24	30	16-14	4.0
25	36	12-11	1.44
26	36	16-11	1.93
27	36	16-14	2.78
28	36	16-18	4.94

FIGURE B.4. Continued.

Solution: We first determine the equivalent eccentricity as shown in Fig. B.3*c*. Thus,

$$e = \frac{M}{P} = \frac{200 \times 12}{400} = 6 \text{ in.}$$

Then, from Fig. B.4*b*, we find that

Minimum size is 24 in. square with 16 No. 10 bars (curve no. 19).

Capacity at 6-in. eccentricity is 410 kips.

Usually, a number of possible combinations of reinforcing bars may be assembled to satisfy the steel area requirement for a given column. Aside from providing for the area, the number of bars must also work reasonably in the layout of the column. Figure B.5 shows a number of tied columns with various number of bars. When a column is small, the preferred choice is usually that of the simple four-bar layout, with one bar in each corner and a single peripheral tie. As the column gets larger, the distance between the corner bars gets larger, and it is best to use more bars so that the reinforcing is spread out around the column periphery. For a symmetrical layout and the simplest of tie layouts, the best choice is for numbers that are multiples of 4, as shown in Fig. B.5a. The number of additional ties required for these layouts depends on the size of the column and the considerations discussed in Sec. B.2.

An unsymmetrical bar arrangement is not necessarily bad, even though the column and its construction details are otherwise not oriented differently on the two axes. In situations where moments may be greater on one axis, the unsymmetrical layout is actually preferred; in fact, the column shape will also be more effective if it is unsymmetrical, as shown for the oblong shapes in Fig. B.5c.

Round columns may be designed and built as spiral columns as described in Sec. B.2, or they may be developed as tied columns with the bars placed in a circle and held by a series of round circumferential ties. Because of the cost of spirals, it is often more economical to use the tied columns; thus they are often used unless the additional strength or other behavioral characteristics of the spiral column are required. In such cases, the column is usually designed as a square column using the square shape that can be included within the round form. It is thus possible to use a four-bar column for small-diameter, round-column forms.

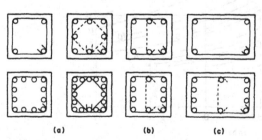

FIGURE B.5. Typical bar placement and tie patterns for rectangular columns.

Figure B.6 gives safe loads for round columns that are designed as tied columns. Load values have been adapted from values determined by strength design methods. The curves in Fig. B.6 are similar to those for the square columns in Fig. B.4, and their use is similar to that demonstrated in Example 1 and 2.

B.4 APPLICATIONS TO MASONRY

The following are some possible applications of the formulas and procedures of the working-stress method to the situations in reinforced masonry.

Flexure

Allowable flexural compression in masonry = F_b, and

$$F_b = 0.333f'_m \quad (UBC)$$

Calculated flexural compression in masonry = f_b, and

$$f_b = \frac{2M}{kjbd^2} \quad \text{(Sec. B.1, formula (2))}$$

Allowable tension in reinforcement = F_s, and

$$F_s = 0.5F_y,$$

or a maximum of 24,000 psi $\quad (UBC)$

Modular ratio $= \dfrac{E_s}{E_m} = \dfrac{30,000}{f'_m}$, or $\dfrac{30,000}{3F_b}$

$$= \frac{10,000}{F_b}$$

(Note: Stresses are in units of ksi.)
Calculated tension in reinforcement = f_s, and

$$f_s = \frac{M}{A_sjd} \quad \text{(Sec. B.1, formula (3))}$$

Resisting moment $= M_R$, and

$$M_R = Kbd^2 \quad \text{(Sec. B.1, formula (8))}$$

$$K = \frac{kjf_b}{2} \quad \text{(Sec. B.1, formula (9))}$$

using K for masonry

A balanced section, with both f_s and f_b at established limiting levels, is defined by a fixed value for k. (See Fig. B.1.) The bal-

anced value for k is

$$k = \frac{1}{1 + f_s/nf_b}$$

$$\text{(Sec. B.1, formula (5))}$$

From geometry alone,

$$j = 1 - \frac{k}{3} \quad \text{(Sec. B.1, formula (6))}$$

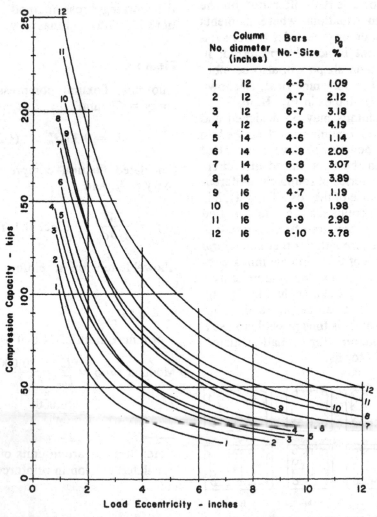

No.	Column diameter (inches)	Bars No. - Size	p_g %
1	12	4-5	1.09
2	12	4-7	2.12
3	12	6-7	3.18
4	12	6-8	4.19
5	14	4-6	1.14
6	14	4-8	2.05
7	14	6-8	3.07
8	14	6-9	3.89
9	16	4-7	1.19
10	16	4-9	1.98
11	16	6-9	2.98
12	16	6-10	3.78

FIGURE B.6. Safe service loads for round tied columns with specified concrete strength of 4 ksi and Grade 60 reinforcement.

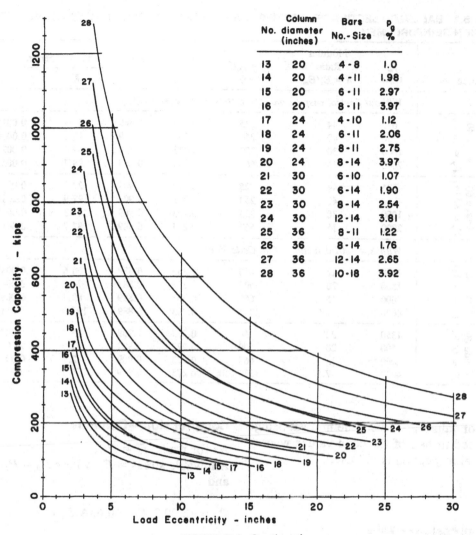

No.	Column diameter (inches)	Bars No.- Size	p_g %
13	20	4 - 8	1.0
14	20	4 - 11	1.98
15	20	6 - 11	2.97
16	20	8 - 11	3.97
17	24	4 - 10	1.12
18	24	6 - 11	2.06
19	24	8 - 11	2.75
20	24	8 - 14	3.97
21	30	6 - 10	1.07
22	30	6 - 14	1.90
23	30	8 - 14	2.54
24	30	12 - 14	3.81
25	36	8 - 11	1.22
26	36	8 - 14	1.76
27	36	12 - 14	2.65
28	36	10 - 18	3.92

FIGURE B.6. Continued.

Since we assume that n is a function of f_b, then k is strictly a function of f_s and f_b or a constant value if they are fixed for a balanced section.

Table B.2 gives values for the various factors just described for the limiting case of a balanced section. Values are given for two common values of f_s, determined by the use of the two common grades of reinforcement: Grade 40 (F_y = 40 ksi) and Grade 60 (F_y = 60 ksi). Values for f'_m in the table relate to some values given in the table in the *UBC* (Table 24-C) for pre-

sumptive design based on specified strengths of masonry units and grades of mortar.

Various design aids can be derived from these relationships. One such aid is described in Sec. C.2 of Appendix C. Some uses of these relationships are demonstrated in the examples in Chapter 5 and 10.

Care must be exercised in the use of code values where special job-site inspection and testing of samples of the construction are required. In Table B.2 two

Table B.2 BALANCED SECTION PROPERTIES FOR RECTANGULAR MASONRY SECTIONS WITH TENSION REINFORCEMENT

Reinforcement	f_m (psi)	Modular Ratio $n = E_s/E_m$	$F_b = f_m/3$ (psi)	Balanced Section Properties			
				k	j	K	$p = A_s/bd$
Without Special Inspection—Code Values Reduced to Half							
Grade 40 $F_y = 40$ ksi	675	44	225	0.333	0.889	33.3	0.00187
	750	40	250	0.333	0.889	37.0	0.00208
	1000	30	333	0.333	0.889	49.3	0.00278
	2000	15	667	0.333	0.889	98.7	0.00556
Grade 60 $F_y = 60$ ksi	675	44	225	0.273	0.909	27.9	0.00128
	750	40	250	0.273	0.909	27.9	0.00142
	1000	30	333	0.273	0.909	41.3	0.00189
	2000	15	667	0.273	0.909	82.7	0.00379
With Special Inspection—Full Code Values							
Grade 40 $F_y = 40$ ksi	1350	22	450	0.333	0.889	66.6	0.00375
	1500	20	500	0.333	0.889	74.0	0.00416
	2000	15	667	0.333	0.889	89.7	0.00556
	4000	7.5	1333	0.333	0.889	197.0	0.01111
Grade 60 $F_y = 60$ ksi	1350	22	450	0.273	0.909	55.8	0.00256
	1500	20	500	0.273	0.909	62.0	0.00284
	2000	15	667	0.273	0.909	82.7	0.00379
	4000	7.5	1333	0.273	0.909	165.4	0.00758

sets of values are given, the first set being reduced to half of the code values where special inspection is not used.

Compression—Walls

Calculated compressive stress = f_a, and

$$f_a = \frac{P}{A_e}$$

where

P = axially applied load
A_e = effective (net) area

Allowable compressive stress = F_a, and

$$F_a = 0.20f'_m\left[1 - \left(\frac{h'}{42t}\right)^3\right] \quad (UBC)$$

Compression—Columns

As for walls, $f_a = P/A_e$.
Allowable stress = F_a, where $F_a = P_a/A_e$, and

$$P_a = (0.20f'_mA_e + 0.65A_sF_{sc})$$
$$\left[1 - \left(\frac{h'}{42t}\right)^3\right]$$

where

A_e = effective (net) area of masonry
A_s = area of steel
F_{sc} = allowable compressive stress in steel = $0.4F_y$ or 24,000 psi maximum

Combined Compression and Bending—Walls and Columns

$$\frac{f_a}{F_a} + \frac{f_b}{F_b} \leqslant 1$$

Appendix C

DESIGN AIDS FOR MASONRY STRUCTURES

This section contains material that supports various common tasks in the investigation and design of elements of masonry structures. The use of these aids is demonstrated in the example problems in the text.

C.1 WALL REINFORCEMENT

Both vertical and horizontal reinforcement in walls are ordinarily achieved with standard deformed steel bars, ranging typically from size No. 3 to size No. 9. Typical spacings follow modules most common in CMU construction. Table C.1 gives values for the rapid determination of alternative choices once a required amount of steel area has been computed. Areas for required reinforcement are ordinarily computed in units of square inches per foot of wall height or length; this is the unit used for the table entries. Table values are given for typical spacings.

Example. Determine the alternative choices for reinforcement for a CMU wall with 8-in.-high block with cells at 8 in. on center along the wall length. Required reinforcement is as follows:

Vertical: $A_s = 0.130$ in.^{2}/ft

Horizontal: $A_s = 0.067$ in.^{2}/ft

Solution: Reading from Table C.1, our choices are as follows:

Vertical: No. 3 at 8 (0.165), No. 4 at 16 (0.150), No. 5 at 24 (0.155), No. 6 at 32 (0.165), No. 7 at 48 (0.150)

Horizontal: No. 3 at 16 (0.082), No. 4 at 32 (0.075), No. 5 at 48 (0.077)

Choices respond to the requirement for an 8-in. modular spacing and a code-required maximum spacing of 48 in. Ordinarily, the usual preference is for the widest allowable spacing, permitting the handling of the fewest number of bars for installation.

Table C.1 AVERAGE REINFORCEMENT PER FOOT IN WALLS (in²/ft)[a]

Bar Spacing (in.)	Bar Size						
	No. 3	No. 4	No. 5	No. 6	No. 7	No. 8	No. 9
6	0.220	0.400	0.620	0.880	1.200	1.580	2.000
8	0.165	0.300	0.465	0.660	0.900	1.185	1.500
12	0.110	0.200	0.310	0.440	0.600	0.790	1.000
16	0.082	0.150	0.232	0.330	0.450	0.592	0.750
18	0.073	0.133	0.207	0.293	0.400	0.527	0.667
24	0.055	0.100	0.155	0.220	0.300	0.395	0.500
30	0.044	0.080	0.124	0.176	0.240	0.316	0.400
32	0.041	0.075	0.116	0.165	0.225	0.296	0.375
36	0.037	0.067	0.103	0.147	0.200	0.263	0.333
40	0.033	0.060	0.093	0.132	0.180	0.237	0.300
42	0.031	0.057	0.088	0.126	0.171	0.226	0.286
48	0.027	0.050	0.077	0.110	0.150	0.197	0.250

[a] Table entry = (bar cross section area)(12)/(spacing).

C.2 FLEXURE IN MASONRY WALLS

Figure C.1 may be used for investigation of bending in masonry walls by the working-stress method. There are three variables in the table, as follows:

The K factor for bending (see Sec. B.4) is plotted vertically on the chart.

The percentage of reinforcement is plotted horizontally on the chart.

The allowable bending stress is represented by the curves on the chart. (Note that the figure uses f_m for the values we give as F_b in this book.)

The chart may be used in a number of ways, a frequent use being the determination of the required reinforcement, as the following example illustrates.

Example 1. A wall of reinforced CMU construction uses 8-in. nominal units ($t = 7.625$ in.) with $f'_m = 1500$ psi. The wall sustains a wind pressure of 20 psf and spans 16.7 ft vertically. Find the minimum reinforcement, considering only the wind loading. Use steel with $F_y = 40,000$ psi.

Solution: From the data we determine the following:

Allowable $F_b = (0.5)(0.33\ f'_m)(1.333)$

$\qquad = (0.5)(0.33)(1500)(1.333)$

$\qquad = 333$ psi

(Assumes 50% reduction for no inspection; one third increase for wind loading.)

Allowable $F_s = 0.5F_y(1.333)$

$\qquad = (0.5)(40,000)(1.333)$

$\qquad = 26,667$ psi

From Table B.2, $n = 40$.
With steel bars in center, $d = t/2 = 3.813$ in.
For the wind load, the wall spans 16.7 ft as a simple span beam. Thus, the maximum bending moment is

$$M = \frac{wL^2}{8} = \frac{(20)(16.7)^2}{8} \times 12$$

$$= 8367 \text{ in.-lb}$$

For the chart

$$K = \frac{M}{bd^2} = \frac{8367}{(12)(3.813)^2} = 48$$

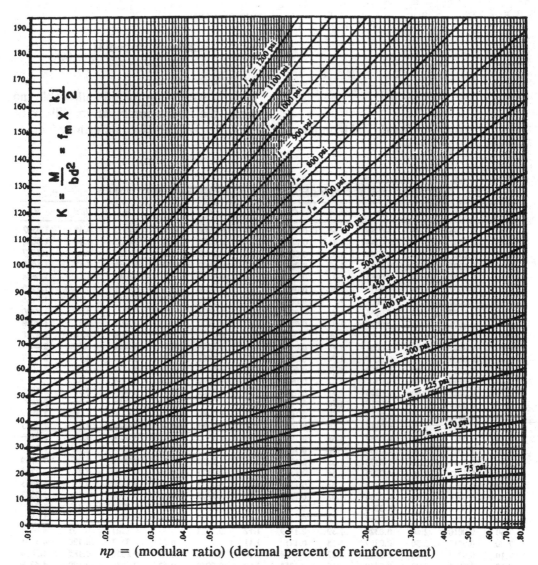

FIGURE C.1. Flexural coefficient *k*-chart for reinforced masonry. Reproduced from *Masonry Design Manual* (Ref. 7) with permission of the publishers, Masonry Institute of America.

Enter the chart (Fig. C.1) at the left with $K = 48$, proceed to the right to intersect a value of $f_m = 333$ psi, then read at the bottom a value of $np = 0.073$, from which

$$p = \frac{0.073}{n} = \frac{0.073}{40} = 0.001825$$

$$A_s = pbd = (0.001825)(12)(3.813)$$

$$= 0.0835 \text{ in.}^2/\text{ft}$$

From Table C.1, possible choice: No. 5 at 40 in., $A_s = 0.093$ in.^{2}/ft.

Since the example does not deal with the problem of combined stress, we consider the following situation.

Example 2. For the wall in Example 1, suppose that the vertical loading combined with the wind loading develops a stress condition such that $f_a/F_a = 0.2$. Then for the combined stress condition,

$$\frac{f_a}{F_a} + \frac{f_b}{F_b} = 1 \quad \text{or} \quad 0.2 + \frac{f_b}{F_b} = 1$$

we can say

Thus,

$$\frac{\text{Loading-induced } M}{\text{Allowable } M} = 0.8$$

$$\frac{f_b}{F_b} = 0.8$$

or, since stress is proportional to moment,

Now, if we design for this allowable moment, the stress combination will be acceptable. We therefore determine a design

Table C.2 STIFFNESS COEFFICIENTS FOR CANTILEVERED MASONRY WALLS

h/d	R_c	h/d	R_c	h/d	R_c	h/d	R_c	h/d	R_c	h/d	R_c	h/d	R_c
9.90	0.0006	5.20	0.0043	1.85	0.0810	1.38	0.1706	0.91	0.4352	0.45	1.4582
9.80	0.0007	5.10	0.0046	1.84	0.0821	1.37	0.1737	0.90	0.4452	0.44	1.5054
9.70	0.0007	5.00	0.0049	1.83	0.0833	1.36	0.1768	0.89	0.4554	0.43	1.5547
9.60	0.0007	4.90	0.0052	1.82	0.0845	1.35	0.1800	0.88	0.4659	0.42	1.6063
9.50	0.0007	4.80	0.0055	1.81	0.0858	1.34	0.1832	0.87	0.4767	0.41	1.6604
9.40	0.0007	4.70	0.0058	1.80	0.0870	1.33	0.1866	0.86	0.4899	0.40	1.7170
9.30	0.0008	4.60	0.0062	1.79	0.0883	1.32	0.1900	0.85	0.4994	0.39	1.7765
9.20	0.0008	4.50	0.0066	1.78	0.0896	1.31	0.1935	0.84	0.5112	0.38	1.8380
9.10	0.0008	4.40	0.0071	1.77	0.0909	1.30	0.1970	0.83	0.5233	0.37	1.9098
9.00	0.0008	4.30	0.0076	1.76	0.0923	1.29	0.2007	0.82	0.5359	0.36	1.9738
8.90	0.0009	4.20	0.0081	1.75	0.0937	1.28	0.2044	0.81	0.5488	0.35	2.0467
8.80	0.0009	4.10	0.0087	1.74	0.0951	1.27	0.2083	0.80	0.5621	0.34	2.1237
8.70	0.0009	4.00	0.0093	1.73	0.0965	1.26	0.2122	0.79	0.5758	0.33	2.2051
8.60	0.0010	3.90	0.0100	1.72	0.0980	1.25	0.2162	0.78	0.5899	0.32	2.2913
8.50	0.0010	3.80	0.0108	1.71	0.0995	1.24	0.2203	0.77	0.6044	0.31	2.3828
8.40	0.0010	3.70	0.0117	1.70	0.1010	1.23	0.2245	0.76	0.6194	0.30	2.4802
8.30	0.0011	3.60	0.0127	1.69	0.1026	1.22	0.2289	0.75	0.6349	0.29	2.5838
8.20	0.0012	3.50	0.0137	1.68	0.1041	1.21	0.2333	0.74	0.6509	0.28	2.6945
8.10	0.0012	3.40	0.0149	1.67	0.1058	1.20	0.2378	0.73	0.6674	0.27	2.8130
8.00	0.0012	3.30	0.0163	1.66	0.1074	1.19	0.2425	0.72	0.6844	0.26	2.9401
7.90	0.0013	3.20	0.0178	1.65	0.1091	1.18	0.2472	0.71	0.7019	0.25	3.0769
7.80	0.0013	3.10	0.0195	1.64	0.1108	1.17	0.2521	0.70	0.7200	0.24	3.2246
7.70	0.0014	3.00	0.0214	1.63	0.1125	1.16	0.2571	0.69	0.7388	0.23	3.3845
7.60	0.0014	2.90	0.0235	1.62	0.1143	1.15	0.2622	0.68	0.7581	0.22	3.5583
7.50	0.0015	2.80	0.0260	1.61	0.1162	1.14	0.2675	0.67	0.7781	0.21	3.7479
7.40	0.0015	2.70	0.0288	1.60	0.1180	1.13	0.2729	0.66	0.7987	0.20	3.9557
7.30	0.0016	2.60	0.0320	1.59	0.1199	1.12	0.2784	0.65	0.8201	0.195	4.0673
7.20	0.0017	2.50	0.0357	1.58	0.1218	1.11	0.2841	0.64	0.8422	0.190	4.1845
7.10	0.0017	2.40	0.0400	1.57	0.1238	1.10	0.2899	0.63	0.8650	0.185	4.3079
7.00	0.0018	2.30	0.0450	1.56	0.1258	1.09	0.2959	0.62	0.8886	0.180	4.4379
6.90	0.0019	2.20	0.0508	1.55	0.1279	1.08	0.3020	0.61	0.9131	0.175	4.5751
6.80	0.0020	2.10	0.0577	1.54	0.1300	1.07	0.3083	0.60	0.9384	0.170	4.7201
6.70	0.0020	2.00	0.0658	1.53	0.1322	1.06	0.3147	0.59	0.9647	0.165	4.8736
6.60	0.0021	1.99	0.0667	1.52	0.1344	1.05	0.3213	0.58	0.9919	0.160	5.0364
6.50	0.0022	1.98	0.0676	1.51	0.1366	1.04	0.3281	0.57	1.0201	0.155	5.2095
6.40	0.0023	1.97	0.0685	1.50	0.1389	1.03	0.3351	0.56	1.0493	0.150	5.3937
6.30	0.0025	1.96	0.0694	1.49	0.1412	1.02	0.3422	0.55	1.0797	0.145	5.5904
6.20	0.0026	1.95	0.0704	1.48	0.1436	1.01	0.3496	0.54	1.1112	0.140	5.8008
6.10	0.0027	1.94	0.0714	1.47	0.1461	1.00	0.3571	0.53	1.1439	0.135	6.0261
6.00	0.0028	1.93	0.0724	1.46	0.1486	0.99	0.3649	0.52	1.1779	0.130	6.2696
5.90	0.0030	1.92	0.0734	1.45	0.1511	0.98	0.3729	0.51	1.2132	0.125	6.5306
5.80	0.0031	1.91	0.0744	1.44	0.1537	0.97	0.3811	0.50	1.2500	0.120	6.8136
5.70	0.0033	1.90	0.0754	1.43	0.1564	0.96	0.3895	0.49	1.2883	0.115	7.1208
5.60	0.0035	1.89	0.0765	1.42	0.1591	0.95	0.3981	0.48	1.3281	0.110	7.4555
5.50	0.0037	1.88	0.0776	1.41	0.1619	0.94	0.4070	0.47	1.3696	0.105	7.8215
5.40	0.0039	1.87	0.0787	1.40	0.1647	0.93	0.4162	0.46	1.4130	0.100	8.2237
5.30	0.0041	1.86	0.0798	1.39	0.1676	0.92	0.4255				

Table C.3 STIFFNESS COEFFICIENTS FOR FIXED MASONRY WALLS

h/d	R_f	h/d	R_f	h/d	R_f	h/d	R_f	h/d	R_f	h/d	R_f	h/d	R_f
9.90	0.0025	5.20	0.0160	1.85	0.2104	1.38	0.3694	0.91	0.7177	0.45	1.736
9.80	0.0026	5.10	0.0169	1.84	0.2128	1.37	0.3742	0.90	0.7291	0.44	1.779
9.70	0.0027	5.00	0.0179	1.83	0.2152	1.36	0.3790	0.89	0.7407	0.43	1.825
9.60	0.0027	4.90	0.0189	1.82	0.2176	1.35	0.3840	0.88	0.7527	0.42	1.874
9.50	0.0028	4.80	0.0200	1.81	0.2201	1.34	0.3890	0.87	0.7649	0.41	1.924
9.40	0.0029	4.70	0.0212	1.80	0.2226	1.33	0.3942	0.86	0.7773	0.40	1.978
9.30	0.0030	4.60	0.0225	1.79	0.2251	1.32	0.3994	0.85	0.7901	0.39	2.034
9.20	0.0031	4.50	0.0239	1.78	0.2277	1.31	0.4047	0.84	0.8031	0.38	2.092
9.10	0.0032	4.40	0.0254	1.77	0.2303	1.30	0.4100	0.83	0.8165	0.37	2.154
9.00	0.0033	4.30	0.0271	1.76	0.2330	1.29	0.4155	0.82	0.8302	0.36	2.219
8.90	0.0034	4.20	0.0288	1.75	0.2356	1.28	0.4211	0.81	0.8442	0.35	2.287
8.80	0.0035	4.10	0.0308	1.74	0.2384	1.27	0.4267	0.80	0.8585	0.34	2.360
8.70	0.0037	4.00	0.0329	1.73	0.2411	1.26	0.4324	0.79	0.873	0.33	2.437
8.60	0.0038	3.90	0.0352	1.72	0.2439	1.25	0.4384	0.78	0.888	0.32	2.518
8.50	0.0039	3.80	0.0377	1.71	0.2468	1.24	0.4443	0.77	0.904	0.31	2.605
8.40	0.0040	3.70	0.0405	1.70	0.2497	1.23	0.4504	0.76	0.920	0.30	2.697
8.30	0.0042	3.60	0.0435	1.69	0.2526	1.22	0.4566	0.75	0.936	0.29	2.795
8.20	0.0043	3.50	0.0468	1.68	0.2556	1.21	0.4628	0.74	0.952	0.28	2.900
8.10	0.0045	3.40	0.0505	1.67	0.2586	1.20	0.4692	0.73	0.969	0.27	3.013
8.00	0.0047	3.30	0.0545	1.66	0.2617	1.19	0.4757	0.72	0.987	0.26	3.135
7.90	0.0048	3.20	0.0590	1.65	0.2648	1.18	0.4823	0.71	1.005	0.25	3.265
7.80	0.0050	3.10	0.0640	1.64	0.2679	1.17	0.4891	0.70	1.023	0.24	3.407
7.70	0.0052	3.00	0.0694	1.63	0.2711	1.16	0.4959	0.69	1.042	0.23	3.560
7.60	0.0054	2.90	0.0756	1.62	0.2744	1.15	0.5029	0.68	1.062	0.22	3.728
7.50	0.0056	2.80	0.0824	1.61	0.2777	1.14	0.5100	0.67	1.082	0.21	3.911
7.40	0.0058	2.70	0.0900	1.60	0.2811	1.13	0.5173	0.66	1.103	0.20	4.112
7.30	0.0061	2.60	0.0985	1.59	0.2844	1.12	0.5247	0.65	1.124	0.195	4.220
7.20	0.0063	2.50	0.1081	1.58	0.2879	1.11	0.5322	0.64	1.146	0.190	4.334
7.10	0.0065	2.40	0.1189	1.57	0.2914	1.10	0.5398	0.63	1.168	0.185	4.454
7.00	0.0069	2.30	0.1311	1.56	0.2949	1.09	0.5476	0.62	1.191	0.180	4.580
6.90	0.0072	2.20	0.1449	1.55	0.2985	1.08	0.5556	0.61	1.216	0.175	4.714
6.80	0.0075	2.10	0.1607	1.54	0.3022	1.07	0.5637	0.60	1.240	0.170	4.855
6.70	0.0078	2.00	0.1786	1.53	0.3059	1.06	0.5719	0.59	1.266	0.165	5.005
6.60	0.0081	1.99	0.1805	1.52	0.3097	1.05	0.5804	0.58	1.292	0.160	5.164
6.50	0.0085	1.98	0.1824	1.51	0.3136	1.04	0.5889	0.57	1.319	0.155	5.334
6.40	0.0089	1.97	0.1844	1.50	0.3175	1.03	0.5977	0.56	1.347	0.150	5.514
6.30	0.0093	1.96	0.1864	1.49	0.3214	1.02	0.6066	0.55	1.376	0.145	5.707
6.20	0.0097	1.95	0.1885	1.48	0.3245	1.01	0.6157	0.54	1.407	0.140	5.914
6.10	0.0102	1.94	0.1905	1.47	0.3295	1.00	0.6250	0.53	1.438	0.135	6.136
6.00	0.0107	1.93	0.1926	1.46	0.3337	0.99	0.6344	0.52	1.470	0.130	6.374
5.90	0.0112	1.92	0.1947	1.45	0.3379	0.98	0.6441	0.51	1.504	0.125	6.632
5.80	0.0118	1.91	0.1969	1.44	0.3422	0.97	0.6540	0.50	1.539	0.120	6.911
5.70	0.0124	1.90	0.1991	1.43	0.3465	0.96	0.6641	0.49	1.575	0.115	7.215
5.60	0.0130	1.89	0.2013	1.42	0.3510	0.95	0.6743	0.48	1.612	0.110	7.545
5.50	0.0137	1.88	0.2035	1.41	0.3555	0.94	0.6848	0.47	1.651	0.105	7.908
5.40	0.0144	1.87	0.2058	1.40	0.3600	0.93	0.6955	0.46	1.692	0.100	8.306
5.30	0.0152	1.86	0.2081	1.39	0.3647	0.92	0.7065				

moment as

$$M = \frac{8367}{0.8} = 10,459 \text{ in.-lb}, \quad \text{or}$$

$$K = \frac{48}{0.8} = 60$$

Using this K on Fig. C.1, we find $np = 0.155$. Then $p = 0.003875$, $A_s = 0.177$ in.^{2}, and No. 6 at 24 in. (0.220 in.^{2}/ft) is the new required reinforcement.

C.3 STIFFNESS FACTORS FOR MASONRY PIERS

Tables C.2 and C.3 present factors that may be used to determine the relative stiffness of masonry piers. The general use of these factors is for the determination of lateral load distribution to piers of different stiffness that share a single lateral load, as discussed in Sec. 5.8.

STUDY AIDS

This section provides the reader with some means to measure his or her comprehension and skill development with regard to the book presentations. When a chapter is completed, the reader should study this section to find out what has been accomplished. Answers to the questions and problems are given at the end of this section.

WORDS AND TERMS

Using the glossary, the index, and the text for each chapter indicated, review the meaning of the following terms. Indicate the significance as well as the simple definitions.

Chapter 1

Masonry
Structural masonry
Mortar
CMU

Reinforced masonry
Unreinforced masonry
Reinforcement
Lintel
Pilaster
Building code

Chapter 2

Masonry unit
Course
Wythe
Solid masonry construction
Cavity wall
Grouted cavity wall
Header
Joint reinforcing
Fired clay
Types of mortar: M, S, N
Standard deformed steel reinforcing bars (rebars)
Control joint
Furred-out space

Chapter 3

Masonry veneer
Minimal construction
Specified compressive strength, f'_m
Allowable stress
Reinforcement—general meaning

Chapter 4

Adobe
Fired brick
Facing brick
Building brick
Brick face
Stretcher
Header
Bonding
Rowlock
Soldier
Patterns: running bond, common bond, English bond, Flemish bond, stack bond
Tooled mortar joint
Pier
Column
Pedestal
Cross-sectional area: gross, net
Walls: bearing, shear, freestanding, retaining, grade
Slenderness (wall or column)
Bearing footing
Fire-cut end (beam)

Chapter 5

Bond beam
Shear key (retaining wall)
Overturn (shear wall, retaining wall)
Diaphragm action
Box system
Tiedown
Relative stiffness of piers

Chapter 6

Rubble stonework
Ashlar stonework
Coursed stonework
Random stonework

Chapter 7

Adobe brick
Rammed-earth construction
Glass block
Clay tile
Architectural terra-cotta

Chapter 8

Thermal bridge
Daily temperature swing
Thermal inertial effect (of building mass)

Chapter 9

Dead load
Live load
Ponding
Load duration
Live-load reduction
Load combinations
Lateral load
Basic wind speed
Wind stagnation pressure
Design wind pressure
Projected area method (wind)
Normal-force method (wind)
Drift
System integration

Chapter 10

Core-bracing system
Perimeter bracing system

Appendix A

Column slenderness (re: fundamental behavior)

Short column

Tall column

Euler buckling curve

Effective buckling length

Column interaction

P-delta effect

Equivalent eccentric compressive force

Kern limit

Cracked section

Pressure-wedge method

Composite element

Appendix B

Effective depth of tension-reinforced member

Balanced section

Tied column

Spiral column

Cover

Modular ratio

GENERAL QUESTIONS

Chapter 1

1. How is the structural development of present construction with masonry exteriors different from that of ancient times?

2. What are some of the nonstructural properties of masonry that make it popular?

3. What is structural masonry?

4. What primary construction element (building part) is the major item produced with structural masonry in present times?

5. Ancient masonry construction was mostly achieved with stone. What are the primary masonry units used for structural masonry in present times?

6. What is generally meant by the use of the term *reinforced masonry*?

7. What is the primary reason for the profusion and diversity of organizations in the masonry industry?

8. Local building codes differ in their criteria for masonry construction; still, they tend to be more similar than different. What are some factors that influence the differences and the basic similarity?

Chapter 2

1. What is a masonry unit?

2. What form of masonry unit is used most extensively for structural masonry in the United States?

3. What is a multiwythe wall?

4. What basic structural function is served by a header unit in masonry construction?

5. What is the architectural significance of the face of a masonry unit?

6. Different forms of units are usually used to produce reinforced and unreinforced construction with CMUs. What are some reasons for this?

Chapter 3

1. Masonry not intended for structural purposes sometimes performs structural tasks. What is a basic reason for this?

2. What is meant by the term *minimal construction*?

3. In what basic type of stress is the strength of masonry usually measured?

4. In addition to the insertion of steel rods, what are some means for reinforcing masonry walls?

5. Why is the quality of the masonry work itself of somewhat higher concern with unreinforced construction, as opposed to reinforced construction?

6. In reinforced masonry construction with hollow units of precast concrete, what basic form of structure is produced as a secondary construction within the masonry?

Chapter 4

1. What is the basic intended purpose of a "building brick"?

2. Despite the continuing use of wall face arrangements such as common bond in brick construction, bonding of wythes is now seldom accomplished with masonry units. What is used instead?

3. What is the purpose for tooling mortar joints?

4. What dimensional ratio distinguishes a wall from a pier? a column from a pedestal?

5. For resistance to vertical loads, masonry walls typically act as bearing walls, but often also serve other structural functions. Describe these other functions.

6. What principal concern for a bearing wall is involved in the effort to center the load on the wall?

7. What is the basic purpose of a fire-cut end on a beam that is seated in a pocket in a masonry wall?

8. Pilasters frequently serve to receive concentrated loads on walls. What other functions can they serve for a wall?

9. In addition to a vertical load resistance, what basic form of support is required for an arch?

Chapter 5

1. What are two primary reasons for the current widespread use of masonry constructed with CMUs?

2. What are the two wall face patterns for arrangement of the units that are used mostly with CMUs?

3. What was the original derivation for the nominal dimensions of CMUs for wall construction?

4. What is the difference between the gross cross section and net cross section of a wall? What basic wall properties are affected by the consideration of the net section?

5. With regard to the basic wall construction, what forms of strength increase can be utilized for CMU construction?

6. What is the basic purpose of the larger voids in the CMUs used for reinforced construction?

7. Building code requirements typically establish a minimum form of construction for reinforced masonry with CMUs. What can be done to increase wall strength beyond this minimum level of structure?

8. How does a surcharge load affect a basement wall?

9. If a single vertical reinforcing bar is used in the voids in a basement wall with CMUs, what is the ideal position for the bar?

10. Other than leaking into the basement space, what is the undesirable effect of a buildup of water in the soil outside a basement wall?

11. What is the basic function of a dropped shear key in the footing for a cantilever retaining wall?

12. What is the significance of keeping the load resultant inside the kern of the footing bottom in a cantilever retaining wall?

13. What is the primary form of structural resistance developed by a diaphragm bracing element?

14. What means are generally used to resist overturn of a shear wall?

15. When a number of masonry wall piers share a lateral load, what is the significance of the relative stiffness of the piers?

16. What is the basis for limitation of the maximum height of a pedestal? for the minimum height?

Chapter 6

1. What is the generally preferred shape for stones in rubble stone masonry?

2. What distinguishes coursed stone masonry from random stone masonry?

3. Why should the mortar not be used to achieve basic stability in stone masonry?

Chapter 7

1. What is ordinarily used for mortar in traditional adobe construction?

2. What is the usual consideration for the width of adobe bricks?

Chapter 8

1. Basic forms of masonry construction are strongly associated with specific regional areas in the United States. What are some reasons for this?

2. What are the typical major influences on the choice of a particular form of masonry construction for a building?

3. For what climate condition is insulation imperative for an exterior masonry wall?

4. What thermal enhancement is served by a cavity in masonry wall construction?

5. What is the potential usefulness of the mass of a masonry wall for control of interior thermal conditions in a building?

Chapter 9

1. What is the reason for tabulating dead and live loads separately for structural design?

2. What is the usual basis for determination of live-load reduction?

3. Why are the lateral (horizontal) effects, rather than the vertical effects, of wind and earthquakes generally more critical for building design?

4. Why are wind surface pressures greater on tall buildings?

5. What basic dimensional constraints must ordinarily be accepted when planning buildings with CMU construction?

EXERCISE PROBLEMS

Chapter 4

1. A brick wall of solid construction is 13 in. thick and has an unbraced height of 16 ft. The construction is solid, unreinforced, and $f'_m = 2000$ psi. The wall sustains a uniformly distributed load of 4000 lb/ft, and the average weight of the material is 145 lb/ft^3. Investigate the wall for compression.

2. Assume the wall in Problem 1 sustains a concentrated load of 30,000 lb from the end of a beam resting on a bearing plate 10 by 18 in. in plan dimensions and centered on the wall. Investigate the wall for bearing and compression.

3. Assume the wall in Problem 1 sustains the given vertical load plus a wind pressure of 35 psf on the outside surface. The wall spans vertically as a simple beam. Investigate the combined load condition.

4. Assume that the load in Problem 2 is placed so that it is 1.5 in. from the center of the wall. Investigate the combined compression and bending condition.

5. If steel reinforcement with allowable stress of 20 ksi is placed in the center of the wall in Problem 4 to develop the tension resistance to the bending, find the total cross-sectional area required for the bars.

Chapter 5

1. A masonry wall of CMUs is 14 ft high and sustains a uniformly distributed load of 4000 lb/ft. Units are 10 in. nominal thickness, have $f'_m = 1500$ psi, and are laid with Type S mortar. Investigate the wall for compressive stress. Assume a density of 100 lb/ft³ for the concrete blocks and an average of 50% void.

2. Assume that the wall in Problem 1 sustains a concentrated load of 20,000 lb at 6-ft centers, delivered to the wall center through a bearing plate that is 8 by 14 in. Investigate the wall for bearing and compression as an unreinforced wall.

3. Investigate the wall in Problem 2 for the combined compression and bending caused by the placement of the load at 3 in. from the wall center.

4. Assume the wall in Problem 3 is built as reinforced construction using steel with allowable $f_s = 20$ ksi.

Three block voids at the location of the load will be grouted, and the reinforcement will be placed in the center of the wall. Determine the required area of steel and investigate for compression in the masonry.

5. Assume the wall in Problem 1 to sustain the given vertical load plus a wind pressure of 25 psf. Investigate the wall
   a. As an unreinforced wall.
   b. As a reinforced wall with minimum vertical reinforcement.

6. A basement wall is to be built as shown in Fig. 5.7 using 12-in. blocks of 140 lb/ft³ density, $f'_m = 1500$ psi, and 50% void. The wall is 12 ft high and sustains soil pressure of 35 psf. Investigate the wall for reinforced construction.

7. A short cantilever retaining wall is to be built as shown in Fig. 5.10, using 10-in. CMUs of 140 lb/ft³ density, $f'_m = 1500$ psi, and 50% void. All voids are grouted, and soil values are as shown in the figure. Dimensions are $H = 5$ ft 8 in., $w = 52$ in., $h = 15$ in., $A = 16$ in. Find the required reinforcement and investigate for the maximum soil pressure.

8. A wall similar to that shown in Fig. 5.17 sustains a lateral load as a shear wall. Find the percentage of the total load resisted by each pier, using the following pier widths: pier 1 = 6 ft, pier 2 = 8 ft, pier 3 = 12 ft, pier 4 = 10 ft. All piers are 10 ft high.

9. A short column of the form shown in Fig. 5.18b is formed of fully grouted 12-in. nominal CMUs (8 by 12 by 16 in.) with $f'_m = 1350$ psi and is laid with Type N mortar. Find the capacity for the column without reinforcement.

10. Assume the column in Problem 9 is reinforced with four No. 7 bars with $F_y = 60$ ksi (allowable $f_s = 24$ ksi) and sustains a 40,000-lb load at 3 in. from the column center. Investigate the column for the combined load, using the approximate method shown in Sec. 5.10, Example 2. The column is 14 ft high.

ANSWERS TO GENERAL QUESTIONS

Chapter 1

1. Today most buildings with masonry materials on the exterior do not have masonry structures but rather veneered construction over a framed supporting structure.
2. Resistance to weather exposure, fire, wear, rot, vermin, and insects. Good acoustic separation. Solid, nondeforming structural support.
3. Usually refers to masonry used to support other construction, such as roofs, floors, or walls above.
4. Walls; mostly of CMUs, brick, or clay tile.
5. CMUs, brick, and clay tile.
6. Hollow construction with two-way steel rebars embedded in concrete poured into the voids, emulating reinforced concrete construction.
7. Regional differences in material usage, performance requirements, and construction practices.
8. Weather conditions and regional concerns for windstorms or earthquakes create differences. Similarity of basic materials, industry standards, and engineering design as well as the nonregional nature of fire, water, and gravity create similarities.

Chapter 2

1. Any individual element capable of being laid in mortar to produce masonry construction.
2. Hollow concrete blocks, called hollow concrete masonry units or CMUs.
3. A wall with more than one vertical layer of masonry units.
4. The tying together of separate wythes in a multiwythe wall.
5. It is the side basically intended for exposure to view in the face of the wall surface.
6. Units with thinner shells and fewer, larger voids permit creation of larger reinforced concrete members in the voids. Conversely, thicker shells and more cross webs (with more voids per unit) make stronger units, where no reinforcing or grouting is used and the construction relies entirely on the strength of the masonry units.

Chapter 3

1. Minimum construction requirements for any "real" masonry result in construction with considerable structural potential. If placed in particular arrangements in the building, this structure may become a real supporting or bracing element, due especially to its relative stiffness.
2. Construction that is produced with marginal adequacy in terms of code requirements; that is, it just satisfies the code.
3. Compressive stress due to direct axial compression force.
4. Using stronger units, using a higher-quality mortar, using a pattern of units with better structural integrity (running bond versus stack bond),

using pilasters, reinforcing corners and edges with stronger units, using form variations to enhance stability (curved plan, corners, intersecting walls, etc.).

5. In reinforced construction a major component of the strength comes from the grouted, reinforced voids. In unreinforced construction the masonry is all there is.

6. A reinforced concrete rigid frame.

Chapter 4

1. For use in structural masonry.

2. Steel ties in horizontal mortar joints.

3. To compact the mortar in the joint, especially at the exposed face.

4. Assuming a constant thickness—plan length for a wall and pier; height for a column and pedestal.

5. Spanning vertically to resist lateral forces (wind, seismic, earth); spanning horizontally as beams; bracing against lateral force in their own planes (as shear walls).

6. Avoiding bending due to load eccentricity.

7. To prevent the beam from prying or twisting the wall if the beam fails (as in a fire).

8. To reinforce the wall for spanning (against wind, etc.). To brace a wall to reduce its slenderness.

9. Horizontal resistance to the thrust at the base of the arch.

Chapter 5

1. Relative time and cost of labor for laying bricks. Ease of producing reinforced construction.

2. Running bond and stack bond.

3. Relation to construction with standard lumber (2 × 4, etc.).

4. The gross section is defined by the outer dimensions; subtract the void and you have the net section. Major properties affected are the average weight of the construction and the strength of the construction.

5. Same as those for unreinforced construction (see response to Question 3 in Chapter 3); plus grouting and reinforcing of more than the minimum required amounts.

6. To make placing of reinforcement and grout easier and permit larger reinforced concrete elements to be created.

7. Grout and reinforce more voids or simply use more than the minimum reinforcing.

8. Increases the soil pressure on the outside of the wall.

9. As close as possible to the inside of the wall.

10. Increased pressure on the wall.

11. Increased resistance to horizontal sliding.

12. Avoiding of a "cracked" section; having compressive bearing stress on the entire footing plan surface.

13. Resistance to shear distortion in its own plane.

14. Development of bending strength in the wall and footing, plus the arrangement of the wall and footing profile to develop an adequate dead-weight-resisting moment.

15. Each individual pier will absorb a portion of the total distributed load in proportion to its own stiffness.

16. Maximum height ensures a pure compressive stress response (no column action in general). Minimum height avoids a footing-like response, with bending and shear effects.

Chapter 6

1. Flat or angular, not rounded.
2. The general creation of horizontal layers (courses).
3. The rock pile should be essentially stable on its own. Thick masonry joints tend to shrink and crack and permit movement of the individual units (stones).

Chapter 7

1. The same mud used to produce the bricks.
2. Desired thickness of the wall, preferring single-wythe walls when possible.

Chapter 8

1. Traditions of the craft, possibly deriving from the heritage of the settlers carried from their native lands. Plus some real responses to conditions; see response to Question 8 of Chapter 1.
2. Local building code requirements; availability of materials and crafts; building form and size; basic structural requirements.
3. Cold climates, where extreme differences in outdoor and indoor temperatures endure for long periods.
4. Creation of a dead-air space and a thermal break between the exterior and interior surfaces.
5. As a thermal intertial mass, holding heat (or cool) as a storage unit and releasing it radiantly to keep the adjoining interior air at a level temperature.

Chapter 9

1. To be able to use them separately in various load combinations for design investigations.

2. The total amount of loaded area (roof or floor surface) supported by a structural member.
3. They tend to be more destabilizing. In addition, most structures are fundamentally designed for vertical gravity loads already.
4. The effects of ground surface drag and sheltering by surrounding structures or ground forms are diminished at higher elevations above the ground surface.
5. Dimensions of the construction affected by the modular dimensions of the CMUs, such as wall thickness, wall height, wall plan lengths, and locations of edges of openings.

ANSWERS TO EXERCISE PROBLEMS

Chapter 4

1. Average compression is 41.8 psi. From Table 4.3, with Type S mortar, allowable is 115 psi.
2. Bearing stress is 167 psi; allowable is 252 psi. Average compression is 49.1 psi; allowable is 115 psi.
3. Bending stress is 39.8 psi. Maximum combined stress is 81.6 psi, not critical. Minimum combined stress is still not a tension stress.
4. Bending stress is 22.8 psi; combined stress is not critical.
5. Required area of reinforcement is 0.385 in.^{2}; could use two No. 4 bars.

Chapter 5

1. Actual compression is 81.7 psi; allowable is 150 psi.
2. Actual compression is 92.5 psi; allowable is 150 psi. Bearing stress is 179 psi; allowable is 390 psi.
3. Combined stress is 198 psi; exceeds maximum of 150 psi.

4. Minimum reinforcement for bending is three No. 9 bars. Maximum bending capacity based on compression (Kbd^2) is 86,800 in.-lb. Works for combined condition using $f_a/F_a + f_b/F_b = 1$ if all voids are filled.

5. a. Combined stress is less than allowable, so wall is OK without reinforcement.

   b. Minimum reinforcement is No. 5 at 40 in., just enough for bending if placed in center of wall.

6. Use No. 6 at 16 or No. 7 at 24. OK for combined stress if all voids filled.

7. Use No. 5 at 32 in wall. Maximum soil pressure is 665 psf.

8. Distribution: pier 1, 12%; pier 2, 21%; pier 3, 38%; pier 4, 29%.

9. Column approximately 28 in.^{2} in section. Capacity is 118 kips.

10. Column is OK; $f_a/F_a + f_b/F_b = 0.55$.

GLOSSARY

This glossary contains a number of entries that relate to masonry and concrete construction. It also contains some of the general vocabulary of structural investigation and design as it has been used in this book.

Absorption. The amount of water that a masonry unit will absorb when immersed in water at room temperature for a stated length of time. It is expressed as a percent of the weight of the dry unit or in pounds of water per cubic foot of net volume for concrete blocks.

Adhesion bond. The adhesion of the mortar and grout to the masonry units (an important structural requirement).

Admixtures. Materials added to cement, aggregate, and water, such as water repellents, air-entraining, or plasticizing aids, coloring agents, or aids to retard or speed up initial hardening (set).

Adobe. Masonry construction that utilizes unburned (not fired) clay units.

Aggregate. Inert particles such as sand, gravel, or rock that when bound together with portland cement and water form concrete.

Anchor bolt. A bolt embedded in concrete or masonry for the purpose of fastening something.

Anchor ties. Any type of fastener used to secure masonry to some stable object, such as another wall; usually used for tension value.

Anchorage. Refers to attachment for resistance to movement; usually a result of uplift, overturn, sliding, or horizontal separation. Tiedown, or holddown, refers to anchorage against uplift or overturn. Positive anchorage generally refers to direct fastening that does not easily loosen.

Area, gross cross-sectional. The area delineated by the out-to-out dimensions of masonry in the plane under consideration.

Area, net cross-sectional. The area of masonry units, grout, and mortar crossed by the plane under consideration, based on out-to-out dimensions.

ASTM. American Society for Testing and Materials.

Backfill. Earth or earthen material used to fill the excavation around a foundation: the act of filling around a foundation.

Bearing foundation. Foundation that transfers loads to soil by direct vertical contact pressure. Usually refers to a shallow bearing foundation—that is, a foundation that is placed directly beneath the lowest part of the building. See also Footing.

Bearing wall. Any masonry wall that supports more than 200 lb per lineal foot superimposed load, or any such wall supporting its own weight for more than one story.

Bed joint. The horizontal layer of mortar on or in which a masonry unit is laid.

Bent. A planar framework, or some portion of one, that is designed for resistance to both vertical and horizontal forces in the plane of the frame.

BOCA. Building Officials and Code Administrators International, Inc., an organization that publishes a model building code (Ref. 3).

Bond:

Common bond. Units laid so that they lap half over each other in successive courses. Also called half-bond.

Mechanical bond. Units laid so that they lap over each other in successive courses. Includes quarter-bond, third bond, and half-or common bond.

Running bond. Lapping of units in successive courses so that the vertical head joints lap. Placing vertical mortar joints centered over the unit below is called center bond, or half-bond, whereas lapping ⅓ or ¼ is called third-bond or quarter-bond.

Stack bond. A bonding pattern where no unit overlaps either the one above or below and all head joints form a continuous vertical line. Also called plumb joint bond, straight stack, jack bond, jack on jack, and checkerboard bond.

Bond beam. One or more courses of masonry units poured solid and reinforced with longitudinal steel bars.

Brittle fracture. Sudden, ultimate failure in tension or shear. The basic structural behavior of so-called brittle materials.

Buckling. Collapse, in the form of sudden sideways deflection, of a slender element subjected to compression.

Cap. Masonry units laid on top of a finished masonry wall or pier. Metal caps are units formed of metal, as for flashing.

Cavity wall. A wall built of two or more wythes of masonry units so arranged as to provide a continuous airspace within the wall. The facing and backing, outer wythes, are tied together with noncorrosive ties (e.g., brick or wire).

Cell (core). The molded open space in a concrete masonry unit.

Centroid. The geometric center of an object, usually analogous to the center of gravity. The point at which the entire mass of the object may be considered to be concentrated when considering moment of the mass.

Chase. A continuous recess built into a wall to receive pipes, ducts, and similar items.

Cleanout. An inspection hole at the base of a cell, used to clean out debris and inspect steel placement. Minimum size shall measure not less than 2″ × 3″.

Column. An isolated vertical member whose horizontal dimension measured at right angles to the thickness does not exceed three times its thickness and whose height is at least three times its thickness.

Composite action. Transfer of stress between components of a member designed so that in resisting loads the combined components act together as a single member.

Composite wall. Reinforced grouted masonry wall, in which inner and outer wythes are dissimilar materials (i.e., block and brick, block and glazed structural units, etc.).

Concrete masonry unit (CMU):

A-block. A two-cell, hollow unit with one end closed and the other end open; also called open-end block.

Bond beam block. A hollow unit with portions depressed $1\frac{1}{4}$ in. or more to permit the forming of a continuous channel for horizontal reinforcing steel and grout.

Channel block. A hollow unit with portions depressed less than $1\frac{1}{4}$ in. to permit the forming of a continuous channel for reinforcing steel and grout.

Concrete block. A hollow concrete masonry unit made from portland cement and suitable aggregates, with or without the inclusion of other materials.

Concrete brick. A solid concrete masonry unit made from portland cement and suitable aggregates, with or without the inclusion of other materials.

H-block. A hollow unit with both ends open; also called double open end.

Offset block. A unit that is not rectangular in shape.

Open-end block. A term applied to both H-blocks and A-blocks.

Pilaster block. Concrete masonry units designed for use in construction of plain or reinforced concrete masonry, pilasters, and columns.

Return L block. Concrete masonry unit designed for use in corner construction for various wall thicknesses.

Sash block. Concrete masonry unit that has an end slot for use in openings to receive metal window frames and premolded expansion joint material.

Scored block. Block with grooves to provide patterns, for example, to simulate raked joints.

Sculptured block. Block with specially formed surfaces in the manner of sculpturing.

Shadow block. Block with face formed in planes to develop surface patterns.

Sill block. A solid concrete masonry unit used for sills or openings.

Slump block. Concrete masonry units produced to "slump" or sag in irregular fashion before they harden.

Solid unit. Refers to masonry units in which the vertical cores are less than 25% of the cross-sectional area.

Split face block. Concrete masonry units with one or more faces having a fractured surface for use in masonry wall construction.

Connector. A mechanical device for securing two or more pieces, parts, or members together, including anchors, wall ties, and fasteners.

Continuity. Most often used to describe structures or parts of structures that have behavior characteristics influenced by the monolithic, continuous nature of adjacent elements, such as continuous vertical multistory columns, continuous multispan beams, and rigid frames.

Control joint. An intentional linear discontinuity in a structure or component designed to form a plane of weakness where cracking can occur in response to various forces so as to minimize or eliminate cracking elsewhere in the structure.

Coping. The material or units used to form a cap or finish on top of a wall, pier, or pilaster.

Corbel. A shelf or ledge formed by projecting successive courses of masonry out from the face of the wall.

Core. See Cell.

Course. A continuous horizontal layer of masonry.

Creep. Plastic deformation that proceeds with time when certain materials, such as concrete and lead, are subjected to constant, long-duration stress.

Curing. The maintenance of proper conditions of moisture and temperature during initial set to develop required strength and reduce shrinkage in concrete products.

Curtain wall. An exterior building wall that is supported entirely by the frame of the building rather than being self-supporting or load-bearing.

Dead load. See Load.

Deep foundation. Foundation used to achieve a considerable extension of the bearing effect of a supported structure below the ground surface. Elements most commonly used are piles or piers.

Diaphragm. A surface element (deck, wall, etc.) used to resist forces in its own plane by spanning or cantilevering. See also Horizontal Diaphragm and Shear Wall.

Dowel. A short cylindrical rod of wood or steel: a steel reinforcing bar that projects from a foundation to tie it to a column or wall.

Dry joint. Head or bed joint without mortar.

Ductile. Describes the load–strain behavior that results from the plastic yielding of materials or connections. To be significant, the plastic strain prior to failure should be considerably more than the elastic strain up to the point of plastic yield.

Efflorescence. A powdery deposit on the face of a structure of masonry or concrete, caused by the leaching of chemical salts by water migrating from within the structure to the surface.

Expansion joint. A discontinuity extending completely through the foundation, frame, and finishes of a building to allow for gross movement due to thermal stress, material shrinkage, or foundation settlement.

f'_m. The compressive strength of f'_m is computed by dividing the ultimate load by the net area of the masonry used in the construction of the test prisms. The gross area may be used in the determination of f'_m for solid masonry units.

Faced wall. A wall in which the facing and backing are so bonded or otherwise tied as to act as a composite element.

Face brick. A brick selected on the basis of appearance and durability for use in the exposed surface of a wall.

Face shell. The side wall of a hollow concrete masonry unit.

Facing. Any material, forming a load bearing part of a wall and used as a finish surface (veneer takes no load other than its own weight).

Fascia. The flat outside facing member of a cornice.

Fire cut. A sloping end cut on a wood beam or joist where it enters a masonry wall. The purpose of the fire cut is to allow the wood member to rotate out of the wall without prying the wall apart, if the floor or roof structure should burn.

Fire resistive. In the absence of a specific ruling by the authority having jurisdiction, the term *fire resistive* is applied to all building materials that are not combustible in temperatures of ordinary fires and will withstand such fires without serious impairment of their usefulness for at least 1 hour, maybe 2 hours, 3 hours, 4 hours, etc.

Fire wall. Any wall that subdivides a building so as to resist the spread of fire.

Flashing. A thin, continuous sheet of metal, plastic, rubber, or waterproof paper used to prevent the passage of water through a joint in a wall, roof, or chimney.

Footing. A shallow, bearing-type foundation element consisting typically of concrete that is poured directly into an excavation.

Furring strip. A length of wood or metal attached to a masonry or concrete wall to permit the attachment of finish materials to the wall using screws or nails.

Grade. 1. The level of ground surface. 2. Rated quality of material.

Grade beam. A horizontal element in a foundation system that serves some spanning or load-distributing function.

Gross compressive strength. The compressive strength of the units based on the total area as defined in Area, Gross Cross-Sectional; expressed in pounds per square inch (psi).

Gross cross-sectional area. See Area.

Grout. A concrete mixture of sand, pea gravel (usually), water, and sometimes admixture that is poured or pumped into the void spaces in hollow construction. Grout encases the reinforcing steel and adds to the fire rating of a block wall.

Header. A horizontal element over an opening in a wall or at the edge of an opening in a roof or floor.

Head joint. The vertical mortar joint between ends of masonry units, sometimes called the cross joint.

Hollow masonry unit. A masonry unit whose net cross-sectional area in any plane parallel to the bearing surface is less than 75% of its gross cross-sectional area measured in the same plane.

Horizontal diaphragm. See Diaphragm. Usually, a roof or floor deck used as part of the lateral bracing system.

ICBO. International Conference of Building Officials.

Imperviousness. The quality of resisting moisture penetration.

Joint reinforcement. Steel wire, bar, or prefabricated reinforcement placed in mortar bed joints.

Kern limit. Limiting dimension for the eccentricity of a compression force if tension stress is to be avoided.

Lateral. Literally means to the side or from the side. Often used in reference to something that is perpendicular to a major axis or direction. With reference to the vertical direction of the gravity forces, wind, earthquakes, and horizontally directed soil pressures are called lateral effects.

Lateral force. A force acting generally in a horizontal direction, such as wind, earthquake, or soil pressure against a foundation wall.

Lintel. A beam placed over an opening in a wall.

Live load. See Load.

Load: The active force (or combination of forces) exerted on a structure.

Dead load. A permanent load due to gravity, which includes the weight of the structure itself.

Live load. Any load component that is not permanent, including those due to wind, seismic effects, temperature change, or shrinkage, but the term is most often used for gravity loads that are not permanent.

Factored load. The service load multiplied by some increase factor for use in strength design.

Service load. The total load combination that the structure is expected to experience in use.

Load-bearing wall. Any wall that, in addition to supporting its own weight, supports the building above it.

Mason. One who builds with bricks, stones, or concrete masonry units: one who works with concrete.

Masonry. Brickwork, blockwork, and stonework.

Masonry unit. A brick, stone, concrete block, glass block, or hollow clay tile intended to be laid in mortar.

Masonry veneer. A single wythe of masonry usually used as a facing over a frame of wood or metal.

Modular. Conforming to a multiple of a fixed dimension.

Modular dimension. A dimension based on a given module, usually 8 in. in the case of concrete block masonry.

Modular masonry unit. A masonry unit whose actual dimensions are one mortar joint less than the modular dimension; that is $8 \times 8 \times 16$ in. is actually $7\frac{5}{8} \times 7\frac{5}{8} \times 15\frac{5}{8}$ in. to allow for $\frac{3}{8}$-in. joints.

Moisture barrier. A membrane used to prevent the migration of liquid water through a floor or wall.

Mortar: A plastic mixture of cementitious materials, fine aggregate, and water, with or without the inclusion of other specified materials.

Fat mortar. A mortar that tends to be sticky and adheres to the trowel.

Harsh mortar. A mortar that, due to an insufficiency of plasticizing material, is difficult to spread.

Lean Mortar. A mortar that, due to a deficiency of cementitious material, is harsh and difficult to spread.

NCMA. National Concrete Masonry Association.

Net cross-sectional area. See Area.

Nominal dimension. A dimension that may vary from the actual dimension by the thickness of a mortar joint, but not more than $\frac{1}{2}$ in. ($\frac{3}{4}$ to $\frac{7}{8}$ in for some slump units). The actual dimension is either $\frac{3}{8}$ or $\frac{1}{2}$ in. less than nominal in concrete masonry.

Overturn. The toppling, or tipping over, effect of lateral loads.

Parapet. The extension of a wall plane or the roof edge facing above the roof level.

Partition. An interior non-load-bearing wall.

Pedestal. A short pier or upright compression member. It is actually a short column with a ratio of unsupported height to least lateral dimension of three or less.

Permeability. The quality of allowing the passage of fluids.

Pilaster. An integral portion of the wall that projects on one or both sides and acts as a vertical beam, a column, an architectural feature, or any combination thereof.

Pointing. Filling mortar into a joint after the masonry unit is laid.

Pour. To cast concrete: an increment of concrete casting carried out without interruptions.

Precast concrete. Concrete cast and cured in a position other than its final position in the structure.

Prism. Units mortared together, generally in stack bond, forming a wallette or assemblage to simulate "in wall construction," grouted or ungrouted per specification requirements. This is the standard test sample for determination of f_m.

Rebar. Reinforcing steel bars of various sizes and shapes used to strengthen concrete or masonry.

Reinforced hollow concrete masonry. Masonry in which reinforcement is embedded in either mortar or grout.

Retaining wall. A structure used to brace a vertical cut or a change in elevation of the ground surface. The term is usually used to refer to a cantilever retaining wall—that is, a freestanding structure consisting only of a wall and its footing, although basement walls also serve a retaining function.

Rowlock. A brick laid on its face edge so that the normal end of brick is visible in the wall face; frequently spelled rolok; sometimes called bull headers.

Running bond. See Bond.

Section. The two-dimensional profile or area obtained by passing a plane through a form. Cross section implies a section at right angles to another section or to the linear axis of an object.

Seismic. Pertaining to ground shock; usually due to earthquakes.

Shear wall. A vertical diaphragm.

Slenderness. Relative thinness. In structures, the quality of flexibility or lack of buckling resistance is inferred by extreme slenderness.

Span. The distance between supports for a beam, girder, truss, vault, arch, or other horizontal structural device: to carry a load between supports.

Soffit. The underside of a beam, lintel, reveal, or a roof overhang.

Soldier. A stretcher brick set on end with face showing on the wall surface.

Spandrel. That part of a wall between the head of a window and the sill of the window above it.

Specified compressive strength. See f'_m.

Stability. Refers to the inherent capability of a structure to develop force resistance as a property of its form, orientation, articulation of its parts, type of connections, methods of support, and so on. Is not directly related to quantified strength or stiffness, except when the actions involve the buckling of structural elements.

Stack bond. See Bond.

Stretcher. A unit laid with its length horizontal and parallel with the face of a wall.

Stucco. Portland cement plaster used as an exterior cladding or siding material.

Tooling. Compressing and shaping the face of a mortar joint with a special tool other than a trowel; also called jointing.

Tuck pointing. The filling in with fresh mortar of cut-out or defective mortar joints in masonry.

UBC. Uniform Building Code.

Ultimate strength. Usually used to refer to the maximum static force resistance of a structure at the time of failure. This limit is the basis for the so-called strength design methods, as compared to the stress design methods that use some established stress limit, called the design stress, working stress, permissible stress, and so on.

Unreinforced. Constructed without steel reinforcing bars or welded wire fabric.

Veneer. A masonry facing that is attached to the backup but not so bonded as to act with it under load, as opposed to Faced Wall.

Wall: A vertical, planar building element.

Bearing walls. Walls used to carry vertical loads in direction compression.

Foundation walls. Walls that are partly or totally below ground.

Freestanding walls. Walls whose tops are not laterally braced.

Grade walls. Walls used to achieve the transition between the building that is above the ground and the foundations that are below it; grade is used to refer to the level of the ground surface at the edge of the building. (see also Grade Beam).

Retaining walls. Walls that resist horizontal soil pressure.

Shear walls. Walls used to brace the building against horizontal forces due to wind or seismic shock.

Wythe (or Withe). Each vertical section of masonry a single unit in thickness.

BIBLIOGRAPHY

The following list contains materials that have been used as references in the development of various portions of the text. Also included are some widely used publications that serve as general references for design of building structures, although no direct use of materials from them has been made for this book. The numbering system is random and merely serves to simplify referencing by text notation. For some special topics, additional references are also given at the ends of some text sections.

1. *Uniform Building Code*, 1988 ed., International Conference of Building Officials, Whittier, CA, 1988. (Used in this book as a primary building code reference and called simply the *UBC*.)
2. *American National Standard Minimum Design Loads for Buildings and Other Structures*, American National Standards Institute, New York, 1982.
3. *The BOCA Basic National Building Code/ 1984*, 9th ed., Building Officials and Code Administrators International, Country Club Hills, IL, 1984. (Called simply the BOCA Code.)
4. *Building Code Requirements for Masonry Structures* (ACI 530-88) and *Specifications for Masonry Structures* (ACI 530.1-88), in a single publication, American Concrete Institute, Detroit, 1989.
5. *Building Code Requirements for Reinforced Concrete* (ACI 318-88), American Concrete Institute, Detroit, 1988. (Called simply the ACI Code.)
6. *CRSI Handbook*, Concrete Reinforcing Institute, Schaumburg, IL, 1984.
7. *Masonry Design Manual*, 4th ed., Masonry Institute of America, Los Angeles, 1989.
8. *Reinforced Masonry Design*, 3rd ed., R. R. Schneider and W. L. Dickey, Prentice-Hall, Englewood Cliffs, NJ, 1987.
9. *Masonry Design and Detailing*, 2nd ed., C. Beall, Prentice-Hall, New York, 1987.
10. *Structural Details for Masonry Construction*, M. Newman, McGraw-Hill, New York, 1988.

11. *Architectural Graphic Standards,* 8th ed., C. G. Ramsey and H. R. Sleeper, Wiley, New York, 1988.

12. *Reinforced Concrete Fundamentals,* 4th ed., P. Ferguson, Wiley, New York, 1979.

13. *Simplified Engineering for Architects and Builders,* 7th ed., H. Parker, prepared by James Ambrose, Wiley, New York, 1989.

14. *Simplified Design of Building Foundations,* 2nd ed., J. Ambrose, Wiley, New York, 1988.

15. *Simplified Building Design for Wind and Earthquake Forces,* 2nd ed., J. Ambrose and D. Vergun, Wiley, New York, 1990.

16. *Fundamentals of Building Construction: Materials and Methods,* 2nd ed., E. Allen, Wiley, New York, 1989.

INDEX